Editors: K. Krickeberg;
R.C. Lewontin;
J. Neyman; M. Schreiber

Biomathematics

Vol. 1:

Mathematical Topics in Population Genetics
Edited by K. Kojima
55 figures. IX, 400 pages. 1970
Cloth DM 68,—; US $26.20
ISBN 3-540-05054-X

This book is unique in bringing together in one volume many,
if not most, of the mathematical theories of population
genetics presented in the past which are still valid and some
of the current mathematical investigations.

Vol. 2:

E. Batschelet
Introduction to Mathematics for Life Scientists
200 figures. XIV, 495 pages. 1971
Cloth DM 49,—; US $18.90
ISBN 3-540-05522-3

This book introduces the student of biology and medicine to
such topics as sets, real and complex numbers, elementary
functions, differential and integral calculus, differential equations, probability, matrices and vectors.

M. Iosifescu; P. Tautu
Stochastic Processes and Applications in Biology and Medicine

Vol. 3:

Part 1: Theory
331 pages. 1973
Cloth DM 53,—; US $20.50
ISBN 3-540-06270-X

Vol. 4:

Part 2: Models
337 pages. 1973
Cloth DM 53,—; US $20.50
ISBN 3-540-06271-8

Distribution Rights for the Socialist Countries: Romlibri,
Bucharest

This two-volume treatise is intended as an introduction for
mathematicians and biologists with a mathematical background
to the study of stochastic processes and their applications in
medicine and biology. It is both a textbook and a survey of the
most recent developments in this field.

Vol. 5:

A. Jacquard
The Genetic Structure of Populations
Translated by B. Charlesworth; D. Charlesworth
92 figures. Approx. 580 pages. 1974
Cloth DM 96,—; US $37.00
ISBN 3-540-06329-3

Population genetics involves the application of genetic information
to the problems of evolution. Since genetics models based on
probability theory are not too remote from reality, the results
of such modeling are relatively reliable and can make important
contributions to research. This textbook was first published
in French; the English edition has been revised with respect
to its scientific content and instructional method.

Prices are subject to change without notice

Springer-Verlag
Berlin
Heidelberg
New York

Lecture Notes in Biomathematics

Edited by S. Levin, Ithaca

1

Paul Waltman

The University of Iowa, Iowa City, IA/USA

Deterministic Threshold Models in the Theory of Epidemics

Springer-Verlag
Berlin · Heidelberg · New York 1974

AMS Subject Classifications (1970): 92-02, 92 A 15

ISBN-13:978-3-540-06652-1   e-ISBN-13:978-3-642-80820-3
DOI: 10.1007/978-3-642-80820-3

PREFACE

These notes correspond to a set of lectures given at the University of Alberta during the spring semester, 1973. The first four sections present a systematic development of a deterministic, threshold model for the spread of an infection. Section 5 presents some computational results and attempts to tie the model with other mathematics. In each of the last three sections a separate, specialized topic is presented.

The author wishes to thank Professor F. Hoppensteadt for making available preprints of two of his papers and for reading and commenting on a preliminary version of these notes. He also wishes to thank Professor J. Mosevich for providing the graphs in Section 5. The visit at the University of Alberta was a very pleasant one and the author wishes to express his appreciation to Professors S. Ghurye and J. Macki for the invitation to visit there. Finally, thanks are due to the very competent secretarial staff at the University of Alberta for typing the original draft of the lecture notes and to Mrs. Ada Burns of the University of Iowa for her excellent typescript of the final version.

TABLE OF CONTENTS

1. A SIMPLE EPIDEMIC MODEL WITH PERMANENT REMOVAL

The basic problem discussed in these lectures is to describe the spread of an infection within a population. As a canonical example one thinks of a small group of individuals who have a communicable infection being inserted into a large population of individuals capable of "catching" the disease. Then an attempt is made to describe the spread of the infection in the larger group. To do this, certain assumptions are required to describe the characteristics of the disease and the mixing of the population. From these assumptions a mathematical model is formulated. The model is analyzed, and the results of the analysis (hopefully) interpreted in epidemiological terms and thereby insight is gained into the nature of the phenomenon.

In the mathematical theory of epidemics a variety of different approaches are utilized. The models in these lectures are all deterministic rather than stochastic — that is, they use differential equations rather than stochastic processes to describe changes in the population. To introduce the basic ideas we describe first a model due to Kermack and McKendrick (see notes). The population is divided into three disjoint classes of individuals:

(S) The susceptible class, i.e., those individuals who
     are not infective but who are capable of contracting
     the disease and becoming infective,

(I) The infective class, i.e., those individuals who are
     capable of transmitting the disease to others,

(R) The removed class, i.e., those individuals who have
     had the disease and are dead, or have recovered and
     are permanently immune, or are isolated until recov-
     ery and permanent immunity occur.

The spread of the infection is presumed to be governed by the following rules:

> (i) The rate of change in the susceptible population is proportional to the number of contacts between members of classes (S) and (I), where we take the number of contacts to be proportional to the product of the number of members of (S) and the number of members of (I),

> (ii) Individuals are removed from the infectious class (I) at a rate proportional to the size of (I),

> (iii) The population is constant.

Hypothesis (i) is a statement of the law of mass action and assumes uniform mixing of the population — contact depends only on the numbers in each class. This is reasonable if the population consists of students in a school where changing classes, attending athletic events, etc., mix the population. It would not be true in an environment where socio-economic factors have a major influence on contacts.

Hypothesis (ii) states that recovery is equally likely among infectives, and in particular does not take into account the length of time any particular individual has been an infective. This is an attempt to use a "statistical quantity," the proportion recovering, to replace an accounting on an individual basis. Later we will see that (ii) yields the survival probability of an infective (as an infective).

Hypothesis (iii) states that we are considering a closed population, in particular, ignoring births, deaths, immigrations, etc.

The progress of an individual can be schematically indicated by

$$S \to I \to R.$$

We treat the population as a continuum. If we denote by $S(t)$, $I(t)$, $R(t)$ the number of individuals in classes (S), (I), (R) respectively, at time t, (i),(ii),(iii) yield the following set of differential equations:

(1.1)
$$\frac{dS}{dt} = -rSI$$

(1.2)
$$\frac{dI}{dt} = rSI - \gamma I$$

(1.3)
$$\frac{dR}{dt} = \gamma I$$

with initial conditions,

$$S(0) = S_0 > 0, \quad I(0) = I_0 > 0, \quad R(0) = 0.$$

The proportionality constant, $r > 0$, is called the infection rate; $\gamma > 0$, the removal rate; and $\rho = \gamma/r$, the relative removal rate. The total number of individuals in the population is denoted by N, where $N = I_0 + S_0$, the initial number of infectives plus the initial number of susceptibles. Suppose now that the I_0 infectives are inserted into the S_0 susceptibles at time $t = 0$.

We note first that from equation (1.1) it follows that $S(t)$ is monotone decreasing. If (1.2) is written

$$\frac{dI}{dt} = I(rS - \gamma)$$

it is obvious that if $S_0 < \gamma/r$, then $\left.\frac{dI}{dt}\right|_{t=0} < 0$, and since $S(t) \le S_0$, it follows that $I'(t) < 0$ for all t. In this case, the infection "dies out," that is, no epidemic can occur. This is a <u>threshold phenomenon</u> — there is a critical value which the initial susceptible population must exceed for there to be an epidemic (or, viewed another way, the relative removal rate must be sufficiently small to allow the infection to spread).

Since $S(t)$ is nonincreasing and positive, $\lim_{t \to \infty} S(t) = S(\infty)$ exists. Since $R'(t) \ge 0$ and $R(t) \le N$, $\lim_{n \to \infty} R(t) = R(\infty)$ also exists. Since $I(t) = N - R(t) - S(t)$, it follows also that $\lim_{t \to \infty} I(t)$ exists. These limits are important qualities — they tell us "how things turn out eventually." Since it will turn out that $I(\infty) = 0$ in this model, the ratio $\dfrac{R(\infty)}{N}$ is a measure of the intensity of the epidemic (the proportion of the susceptible population which has contracted the disease). To determine these limits we manipulate the equations.

Dividing equation (1.1) by equation (1.3) yields

$$\frac{dS}{dR} = \frac{-r}{\gamma} S$$

or

(1.4)
$$S = S_0 \exp\left(\frac{-r}{\gamma} R\right)$$

$$\geq S_0 \exp \frac{-r}{\gamma} N = \alpha > 0.$$

Thus $S(\infty) > 0$, or there will always be susceptibles remaining in the population. Thus some individuals will escape the disease altogether, and, in particular, the spread of the disease does not stop for lack of a susceptible population.

Some insight into the progress of the epidemic can be seen by examining trajectories in the $S - I$ plane, the phase plane of equations (1.1) and (1.2). First note that critical points lie on the line $I = 0$. The differential equation of the trajectory is (dividing (1.2) by (1.1))

$$\frac{dI}{dS} = -1 + \frac{\gamma}{rS}$$

or

$$I = I_0 - S + S_0 + \frac{\gamma}{r} \log \frac{S}{S_0} = N - S + \rho \log \frac{S}{S_0}.$$

Viewed another way, solution curves in the phase plane are described by

$$\varphi(S,I) = S + I - \rho \log S = C.$$

Sample curves are plotted in Figure 1.1 where we have chosen the scaling so that $N = 1$.

Since $S(t)$ is decreasing, a curve is traversed from right to left. Since the only critical points lie on the line $I = 0$, the curve approaches $(S(\infty), 0)$ as $t \to \infty$. Hence, necessarily $I(\infty) = 0$. The maximum number of infectives occurs at $S = \rho$. The threshold can also be seen from Figure 1.1. If the initial condition (S_0, I_0) falls to the left of $S = \rho$ then no epidemic occurs; $I(t)$ merely goes monotonely to zero. If $S_0 > \rho$ then the number of infectives increases until S passes through ρ, after which the number of infectives falls to zero. Since $I_0 > 0$ it is necessarily the case that $S(\infty) < \rho$.

To determine $S(\infty)$ we note that since $I(\infty) = 0$, $S(\infty) = N - R(\infty)$ or using (1.4), $S(\infty)$ is a root of

(1.5)
$$z = S_0 \exp \frac{-1}{\rho} (N-z).$$

To see that (1.5) does in fact have a positive root we note that if

$$f(z) = S_0 \exp\left\{\frac{-1}{\rho} (N-z)\right\} - z$$

then $f(0) > 0$ and $f(N) < 0$ since $S_0 < N$. Hence there is a positive root. Let z_0 denote a root and observe that

$$f'(z_0) = -1 + \frac{S_0}{\rho^2} \exp - \frac{1}{\rho} (N-z_0)$$

$$= -1 + \frac{z_0}{\rho}.$$

Since $f''(z) > 0$ and $f(N) < 0$ there is exactly one root z_0 and $z_0 < \rho$.

FIGURE 1.1

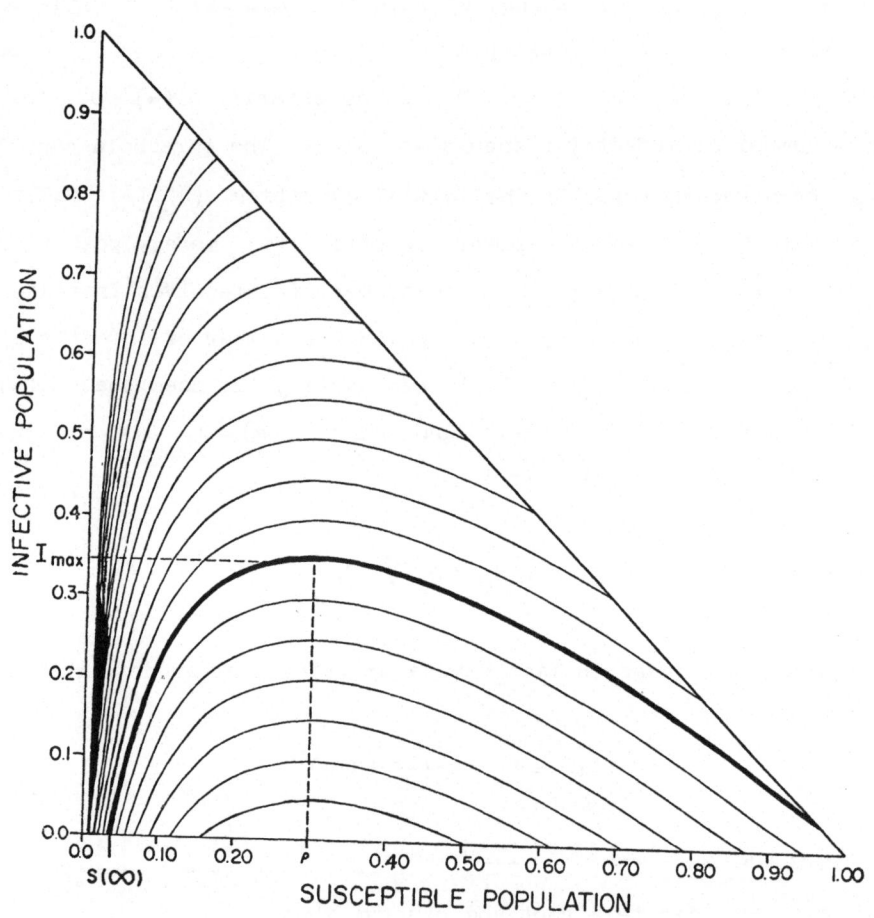

The foregoing is summarized in the following theorem.

THEOREM 1.1. If $S_0 < \rho$, I(t) goes monotonely to zero. If $S_0 > \rho$, the number of infectives increases as t increases and then tends monotonely to zero. $\mathrm{Lim}_{t \to \infty}$ S(t) = S(∞) exists and is the unique root of the transcendental equation

$$S_0 \exp \frac{-1}{\rho} (N-z) - z = 0.$$

Notes:

The model described in this section is that of

> W. D. Kermack and A. G. McKendrick, A Contribution
> to the Mathematical Theory of Epidemics, Journal
> of the Royal Statistical Society, Ser. A, 115
> (1927), pp. 700-721.

In addition to the material presented here, they found approximations to the solution of the equations and compared their predicted results to an actual epidemic, a plague in Bombay in 1905-06. Their curve is presented as Figure 1.2. The ordinate in Figure 1.2 is number of deaths per week and the abscissa is time in weeks. Since almost all cases terminate fatally the ordinate is approximately dR/dt. The curve is generated by

$$\frac{dR}{dt} = 890 \, \mathrm{sech}^2(.2t-3.4).$$

The equations were later solved exactly by Kendall. His solution, as well as the approximate solution of Kermack-McKendrick can be found in

> N. T. J. Bailey, The Mathematical Theory of Epidemics,
> Griffin Book Co., 1957.

This book contains the analysis of many other models, stochastic as well as deterministic, and is an excellent guide to the earlier literature on the subject. Bailey's book has been updated in the following

FIGURE 1.2

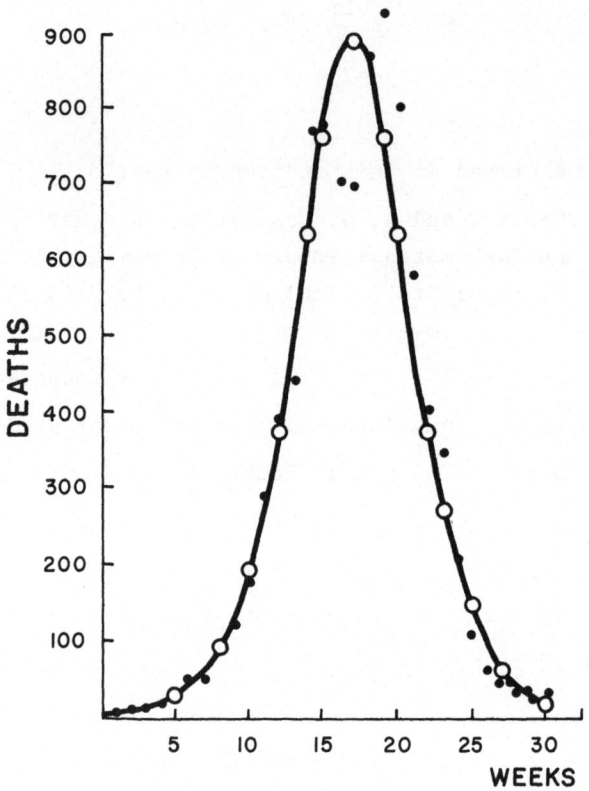

survey article:

K. Dietz, Epidemics and Rumors: A Survey, J. Roy. Statist. Soc. Ser. A 130(1967), 505-528.

The "law of mass action" is discussed in more detail in

E. D. Wilson and J. Worcester, The Law of Mass Action in Epidemiology I & II, Proc. Nat. Acad. Sci. 31(1945), 24-34; 31(1945), 109-116.

2. A MORE GENERAL MODEL AND THE DETERMINATION OF
THE INTENSITY OF AN EPIDEMIC

In the preceding section the ultimate behavior of the model could be determined by finding the roots of a transcendental equation. We explore this idea further with a more general model and procedure. The contents of this chapter are based on a paper of Hoppensteadt (see notes at the end of the section). In order to formulate this more general model, the previous definition of the class (S) is modified to be "susceptible and unexposed to the infection," and a class (E) of "exposed but not yet infective" individuals is added to the model. Schematically, this model is of the form

$$S \to E \to I \to R.$$

We postulate the following:

(i) The rate of exposure of susceptibles to the infection is proportional to the number of contacts between susceptibles and infectives,

(ii) There is given a monotone increasing function $\tau(t), \tau(0) = 0$, $\tau(t) \le t$, such that an individual first exposed at time $\tau(t)$ becomes infective at time t (thus $t - \tau(t)$ is the length of time spent in the class E by an individual who becomes infective at time t),

(iii) There is a given nonnegative, monotone decreasing function $P(\delta)$ which represents the proportion of

those individuals who became infective at time t
who will survive as infectives to $t + \delta$ ($P(\delta)$ is
called the survival probability). Further, we
assume $\int_0^\infty P(x)dx < \infty.$ (If the length of the period
of infectivity of the disease is at most σ, then
$P(x) = 0$, for $x \in (\sigma, \infty)$ but we want to allow the
possibility that $\sigma = +\infty.$),

(iv) The population is constant.

Hypotheses (i) and (iv) are as in the previous model. With
some diseases there is a latent period after exposure to the disease, that is, a period where the individual may be considered to
"have" the disease but is not able to communicate it to others.
This is the class (E). One of the possibilities, and one of the
common assumptions in epidemic models, is that $t - \tau(t)$ is constant. In the previous model, $\tau(t) \equiv t$, or there was no delay
before becoming infective. The point of hypothesis (ii) is to add
this delay phenomenon to the model.

In Hypothesis (iii) the function $P(\delta)$, the survival probability, is fixed instead of prescribing the recovery (or removal) rate.
In Section 1 the recovery rate was assumed to be proportional to the
number of infectives. The length of the infectious period ·is also
now taken into account. The infective period of the disease is distributed over an interval $(0, \sigma)$ — with removal certain after an
individual has been infective for time σ. $P(\delta)$ prescribes the distribution.

From (i) it follows that

$$S' = -rSI$$

(2.1)

$$S(0) = S_0.$$

From (ii) and (iii) it follows that the number of new infectives
introduced into the population at time   t   is given by

$$-\int_{t-\sigma}^{t} \frac{dS(\tau(x))}{dx} P(t-x)H(x)dx$$

where $H(x) = 0$ if $x < 0$ and $H(x) = 1$, $x > 0$.   Thus if $I_0(t)$
describes the number of initial infectives still present as infectives at time   t,   the total number of infectives is described by

$$(2.2) \qquad I(t) = I_0(t) - \int_{t-\sigma}^{t} \frac{dS(\tau(x))}{dx} P(t-x)H(x)dx.$$

The remaining equations of the model are

$$E(t) = S(\tau(t)) - S(t),$$

$$R(t) = N - S(t) - I(t) - E(t).$$

Some additional comment about the initial function $I_0(t)$
seems appropriate. It clearly makes a difference whether the initial infectives, $I_0(0)$ in number, inserted into the population at
time zero, became infective at time $-\sigma + \epsilon$ ($\epsilon > 0$ and small)   and
hence will be removed quickly from the population, or whether they
have just become infective and can be expected to be in the population as infectives for a long time. In the latter case the infection
is much more likely to spread into the susceptible population. Thus
more information is required than just the initial number of infectives. Either $I_0(t)$ must be given as an initial condition or
there are additional assumptions to allow $I_0(t)$ to be computed.
    We show first that the model being discussed includes the model

of the previous section. Suppose that for any infective existing at time x the probability of surviving as an infective to time t is given by $P = e^{-\gamma(t-x)}$ and $\sigma = +\infty$. With $\tau(t) = t$ this will yield the previous model. If I_0 initial infectives are inserted into the population at $t = 0$, then the number of these initial infectives at any future time t is given by $I_0(t) = I_0 e^{-\gamma t}$. Thus equations (2.1) and (2.2) become

$$S' = -rSI, \quad S(0) = 0,$$

$$I(t) = I_0 e^{-\gamma t} + \int_0^t (rIS)(x) e^{-\gamma(t-x)} dx.$$

The second equation may be differentiated to obtain

$$I'(t) = -\gamma I_0 e^{-\gamma t} + r(t)I(t)S(t) - \gamma \int_0^t (rIS)(x) e^{-\gamma(t-x)} dx$$

or

$$I'(t) = rIS - \gamma I.$$

This is the model of Section 1. Note that some infectives may survive for all time $(\sigma = +\infty)$ and that probability of survival as an infective is independent of how long the individual has been infective. Both of these statements are criticisms of the model in Section 1 and future models will attempt to correct these difficulties.

For convenience we suppose that σ is finite and inquire as to the ultimate behavior of the epidemic, that is, we seek the limiting behavior of the functions $S(t)$, $I(t)$, $E(t)$, and $R(t)$. Since $S(t)$ is positive and decreasing, $\lim_{t \to \infty} S(t) = S(\infty)$ exists. Putting equation (2.1) in integrated form and taking a limit as $t \to \infty$ yields

(2.3)
$$S(\infty) = S_0 \exp\left[-r\int_0^\infty I(x)dx\right].$$

If $\tau(\infty) = 0$, then no new infectives can occur and (2.3) becomes

$$S(\infty) = S_0 \exp -r\left(\int_0^\sigma I_0(x)dx\right),$$

and $S(\infty)$ is completely determined. Hence we can assume $\tau(\infty) > 0$. For later use, if $S(\infty) > 0$, we define ν by $S(\tau(\infty)) = e^\nu S(\infty)$. Let

$$\int_0^t I(x)dx = m(t) - n(t)$$

where

$$m(t) = \int_0^t I_0(x)dx.$$

Hence $-n(t)$ is the integral of the number of new susceptibles in the population, and from equation (2.2)

$$n(t) = \int_0^t \int_{u-\sigma}^u \frac{dS(\tau(x))}{dx} P(u-x)H(x)dxdu.$$

An interchange of limits gives

$$n(t) = \int_{-\sigma}^t \int_x^{x+\sigma} \frac{dS(\tau(x))}{dx} P(u-x)H(x)dudx$$

$$= \int_0^t \frac{dS(\tau(x))}{dx} \int_x^{x+\sigma} P(u-x)dudx.$$

Substituting $y = u - x$ and integrating yields

$$n(t) = \left(\int_0^\sigma P(y)dy\right)[S(\tau(t)) - S(0)].$$

On letting $t \to \infty$, and using the definition of ν this becomes

$$n(\infty) = \left(\int_0^\sigma P(y)dy\right)(e^\nu S(\infty) - S_0).$$

Since

$$S(\infty) = S_0 \exp[-r(m(\infty) - n(\infty))],$$

and $m(\infty) = \int_0^\sigma I_0(x)dx,$

(2.4) $\quad S(\infty) = S_0 \exp\left\{-r\left[\int_0^\sigma I_0(x)dx - \left(\int_0^\sigma P(y)dy\right)(e^\nu S(\infty) - S_0)\right]\right\}.$

We check (2.4) against the result (1.5) of the previous section. As noted already in this case $\tau(t) = t$, $\nu = 0$, $\sigma = +\infty$, $I_0(t) = I_0 e^{-\gamma t}$, and $P(\delta) = e^{-\gamma \delta}$. We insert these values in (2.4) (keeping in mind that this is merely formal since in the derivation σ was assumed to be finite) and obtain

$$S(\infty) = S_0 \exp\left\{r\left[\frac{-I_0}{\gamma} + \frac{1}{\gamma}(S(\infty) - S_0)\right]\right\}$$

$$= S_0 \exp\left\{\frac{r}{\gamma}\left[S(\infty) - I_0 - S_0\right]\right\}$$

$$= S_0 \exp\left\{\frac{-1}{\rho}(N - S(\infty))\right\}$$

which agrees with (1.5) of Section 1.

To extend the above to the case $\sigma = +\infty$, we note that all of the foregoing analysis applies except possibly for the interchange of the order of integration in deriving the expression for $n(t)$. $\int_0^\infty P(x)dx$ $< \infty$ is sufficient to insure the applicability of the Fubini Theorem and hence (2.4) holds for $\sigma = +\infty$. Given the initial data, and ν, $S(\infty)$ can be determined by finding the roots of (2.4). This is the principal result of this section.

We want now to manipulate the expression (2.4) into a more useful form. In particular, we want to rewrite it to show epidemic parameters, that is, to bring out certain nondimension parameters which can be interpreted as constants of the initial problem. Since $xP(x) \to 0$ as $x \to \infty$, the "life expectancy" E of an infective (as an infective) is given, after an integration by parts, by

$$E = \int_0^\infty P(x)dx.$$

If

$$\beta = S_0 rE$$

then β measures the expected number of the initially susceptible population who will be exposed to one initial infective. It is a measure of the potential for an epidemic, since if it is less than one, it can be expected that the infection does not spread. (2.4) can be rewritten as

$$e^\nu \frac{S(\infty)}{S_0} = \exp\left\{\nu - r\int_0^\sigma I_0(x)dx + rES_0\left[\frac{e^\nu S(\infty)}{S_0} - 1\right]\right\}.$$

Letting $F = e^\nu \frac{S(\infty)}{S_0}$ and using the definition of β (note that F and β are nondimensional) then this becomes

$$F = \exp\left\{\beta\left(F - 1 + \frac{\nu - r\int_0^\sigma I_0(x)dx}{\beta}\right)\right\}$$

or

(2.5)
$$F = \exp\{\beta(F - 1 - \epsilon)\}$$

where $\epsilon = \dfrac{r\int_0^\sigma I_0(x)dx - \nu}{\beta}$. Equation (2.5) is the "more useful form"

we have been seeking. We show first that it has a root.

If $f(z) = \exp\beta(z - 1 - \epsilon) - z$, then since $f(0) > 0$ and $f(1) < 0$

(if $\epsilon > 0$), there is a root in $(0,1)$. To see that $\epsilon > 0$, we note

that

$$1 > \frac{S(\tau(\infty))}{S_0} = \exp -r(m(\infty) - n(\infty) + \nu) > \exp(-rm(\infty) + \nu)$$

or

$$r\int_0^\sigma I_0(x)dx - \nu > 0. \quad \text{Since} \quad \beta > 0, \quad \epsilon > 0.$$

In case $\beta > 1$, a better upper bound on F can be obtained.

Let $g(x) = x - \dfrac{1}{1+\epsilon}\log ex$. We show that $g(x) > 0$ for $x \geq 1$. First

of all, $g(1) = 1 - \dfrac{1}{1+\epsilon} > 0$, and

$$g'(x) = 1 - \frac{1}{(1+\epsilon)x} > 0$$

since $x > 1$ and $\epsilon > 0$. Hence $e^{-(1+\epsilon)g(x)} < 1$ or $(e^{1-(1+\epsilon)x})x < 1$.

Thus

$$f(1/\beta) = e^{1-(1+\epsilon)\beta} - 1/\beta < 0$$

and there is a root in $(0, 1/\beta)$, that is,

(2.6)
$$0 < F < 1/\beta.$$

Since $f''(z) > 0$, the root is unique.

Graphs of (2.5) are plotted in Figure 2.1. The use of ϵ suggests that it is expected that this term is small. $\int_0^\sigma I_0(x)dx$ is a measure of the amount of infectivity inserted in the population and we have in mind that this is small. Some caution is needed in using this interpretation since our ϵ has a β in the denominator. However, the interesting case is $\beta > 1$ and then "small" initial data does mean small ϵ. Returning to the case in Section 1, $\nu = 0$,

$$\int_0^\sigma I_0(x)dx = \frac{I_0}{\gamma}, \quad \beta = \frac{rS_0}{\gamma} \quad \text{and hence} \quad \epsilon = \frac{I_0}{S_0},\quad \text{the ratio of initial}$$

infectives to initial susceptibles; as noted before, we anticipate I_0 is small compared with S_0.

The quantity $N - e^\nu S(\infty)$ gives the number of initially susceptible individuals who became infective and hence the intensity of the epidemic (which we denote now by $\mathcal{J}$) is given by

$$\mathcal{J} = 1 - \frac{e^\nu S(\infty)}{N}.$$

In terms of the quantity F introduced above, this is

(2.7) $$\mathcal{J} = 1 - \frac{S_0}{N} F.$$

For the given initial data and β, F could be determined by the graph in Figure 2.1 and the intensity calculated by (2.7). Since $\frac{S_0}{N}$ is close to one, $1 - F$ is an obvious first approximation to the intensity.

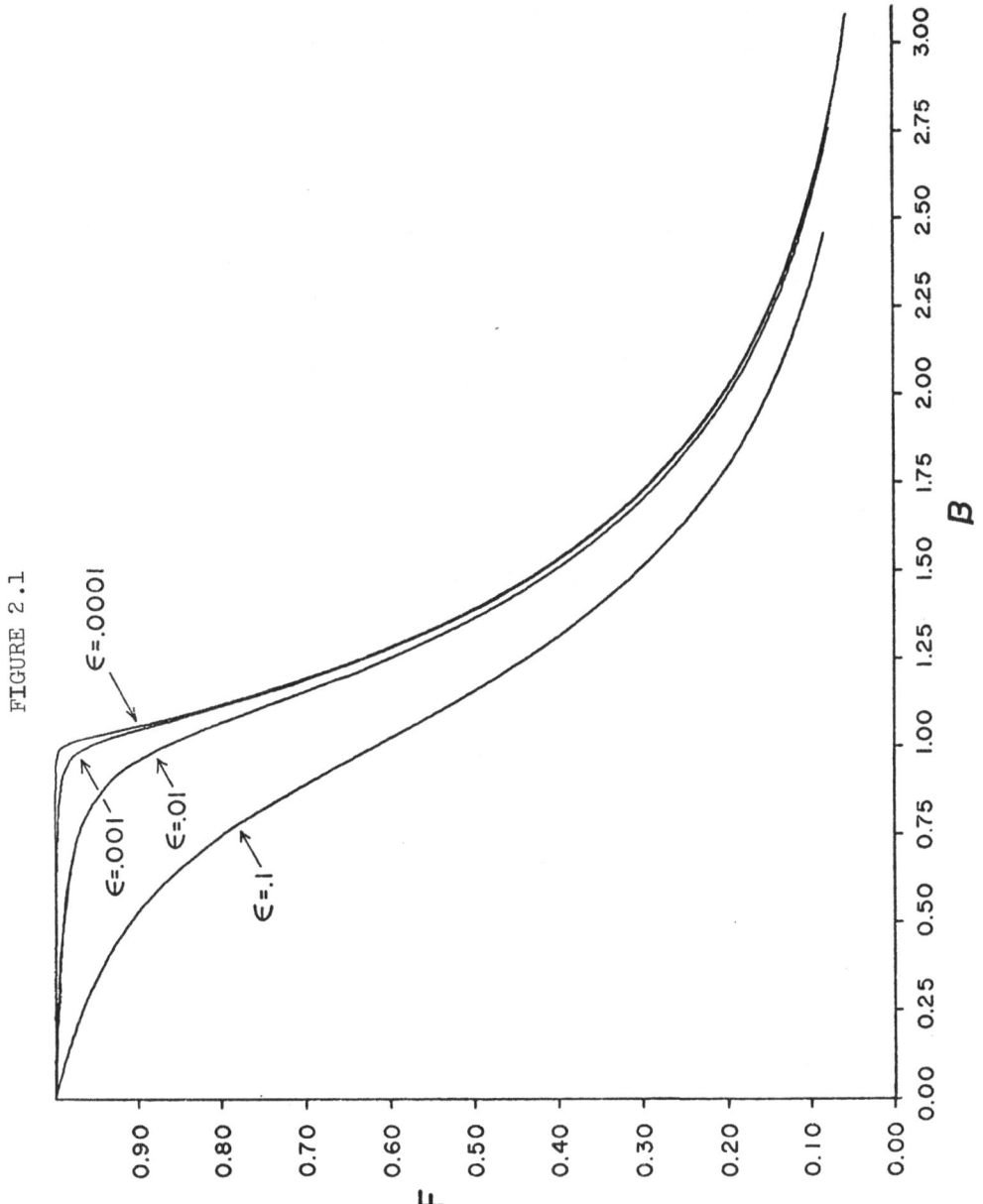

FIGURE 2.1

<u>Notes</u>:

The general procedure in Section 2 is due to

> F. Hoppensteadt, Thresholds for Deterministic Epi-
> demics, (to appear).
> A graph like Figure 2.1 appears there.

The question of existence of a solution of equations (2.1) (2.2)
(and uniqueness and continuous dependence on the data) was not raised.
Techniques which will be developed later for more complex models can
be modified to work for (2.1) (2.2). Basically it involves the use
of a fixed point theorem and a continuation argument.

3. A THRESHOLD MODEL

In the model discussed in the previous section, the function $\tau(t)$ was prescribed a priori. Except for the case of constant delay, it is by no means clear how one would expect to know this function. In this section we add a phenomenon to the model, a threshold effect, which yields the delay as a function of $I(t)$. In order to allow for variations in the infection rate, we now allow r, of the previous models, to be a function $r(t)$. For example, it might be periodic to account for seasonal variations. Specifically, we assume the four classes S, E, I, R are as before and

   (i)   The rate of exposure of susceptibles to infectives
          at time t is given by $-r(t) S(t) I(t)$,

   (ii)   An individual who becomes infective at time t re-
          covers from the infection (is removed from the pop-
          ulation) at time $t + \sigma$, σ a given positive constant,

   (iii) An individual who is first exposed at time τ be-
          comes infective at time t if

$$\int_{\tau}^{t} \rho(x) I(x) dx = m$$

          where $\rho(t)$ is a given positive function and m is a
          given positive constant,

   (iv)   The population is constant.

   (i), except as noted above, that r is now time dependent, and (iv) are as in the previous model. (ii) corresponds to a choice of

$$P(\delta) = \begin{cases} 1, 0 \leq \delta < \sigma \\ 0, \delta \geq \sigma \end{cases}$$ in the model discussed in Section 2. It is in

assumption (iii) that the model radically differs from the previous ones.

A motivation for (iii) is that the human body can often control a small exposure to an infection, that is, there is a tolerance level below which the body's immune system can combat exposure to the infection. When too large an exposure results, the individual contracts the disease. The amount of exposure received depends on the duration of the exposure and the "amount of infectivity" about him. We take the latter to be proportional to the number of infective individuals in the population (where again we allow for seasonal variations). Thus in the interval t, t +h an exposure of

$$\int_t^{t+h} \rho(x)I(x)dx$$

is accumulated where $\rho(x)$ is a proportionality function which is a measure of the amount of infection communicated per infective (virulence). When the total exposure reaches the threshold m, the individual moves from class (E) to class (I).

The threshold does not directly allow for constant delay but (iii) could be modified to be

(iii)′ An individual who is first exposed at time τ becomes infective at time t if

$$\int_\tau^t [\rho_1(x) + \rho_2(x)I(x)]dx = m.$$

If $\rho_2(x) \equiv 0$, $\rho_1(x) = \rho$, a constant, this leads to

$$\tau(t) = t - m/\rho,$$

the case of constant delay. This case will be subsumed in the model of the next section, so only the simpler·threshold will be treated here.

Again we assume I_0 infective individuals are inserted in the population at $t = 0$. As before, the function $I_0(t)$ must be known. It is reasonable to assume that condition (ii) applies to these individuals as well. Then it is sufficient to know the past history of the I_0 individuals inserted at $t = 0$. Of course, in view of (ii), only the past back to time $-\sigma$ is necessary. We assume, as an initial condition a monotone function, $I_0(t)$, $-\sigma \leq t \leq 0$, $I(-\sigma) = 0$, $I_0(0) = I_0$. The "future" of these initial infectives is readily obtainable by applying condition (ii). In fact, it is convenient to extend the function $I_0(t)$ to the real line. Define such an extension by

$$I_0(t) = \begin{cases} 0, & |t| > \sigma \\ I_0(t), & -\sigma < t \leq 0 \\ I_0(0) - I_0(t-\sigma), & 0 \leq t \leq \sigma. \end{cases}$$

The equations for the model then become

$$(3.1) \qquad S'(t) = -r(t)S(t)I(t), \quad S(0) = S_0,$$

$$(3.2) \qquad I(t) = I_0(t) - \int_{t-\sigma}^{t} \frac{d}{dx} S(\tau(x))H(x)dx,$$

$$(3.3) \qquad E(t) = S(\tau(t)) - S(t),$$

$$(3.4) \qquad R(t) = N - S(t) - I(t) - E(t),$$

$$(3.5) \qquad \int_{\tau(t)}^{t} \rho(x)I(x)dx = m,$$

where in (3.2) $H(x) = 0$, $x < 0$, $H(x) = 1$, $x > 0$.

For the infection to spread, some of the initial susceptible population must become infective before time σ. Thus we assume there is a $t_0 < \sigma$ such that

$$\int_0^{t_0} \rho(x)I_0(x)dx = m.$$

We call an initial function $I_0(x)$ which satisfies this, in addition to the above conditions, <u>admissible</u>.

If $\rho(t)$ is continuous and positive, and if a continuous positive solution $I(t)$ exists, then $\tau'(t)$ exists and $\tau(t)$ is monotone increasing. Assuming this, the simple nature of $P(\delta)$ allows equation (3.2) to be rewritten as

$$I(t) = \begin{cases} I_0(t), & 0 \le t \le t_0 \\ I_0(t) + \int_0^{\tau(t)} r(x)S(x)I(x)dx, & t_0 \le t \le \sigma \\ \int_{\tau(t-\sigma)}^{\tau(t)} r(x)S(x)I(x)dx, & t \ge \sigma . \end{cases}$$

If we adopt the convention that $\tau(t) = 0$, $t \le t_0$, this can be rewritten

$$I(t) = I_0(t) + \int_{\tau(t-\sigma)}^{\tau(t)} S'(x)dx$$

or

$$(3.2) \qquad I(t) = I_0(t) + S(\tau(t-\sigma)) - S(\tau(t)).$$

Thus, as an equivalent problem, we seek three functions $I(t) > 0$, $S(t) > 0$, $\tau(t)$ which satisfy

(3.1)
$$S'(t) = -r(t)S(t)I(t), \quad S(0) = S_0$$

(3.2)
$$I(t) = I_0(t) + S(\tau(t-\sigma)) - S(\tau(t))$$

(3.5)
$$\int_{\tau(t)}^{t} r(x)I(x)dx = m, \quad t > t_0$$

$$\tau(t) = 0, \quad t \leq t_0.$$

(If I, S, τ are known, E and R can be found directly from (3.3) and (3.4).)

THEOREM 3.1. If $\sigma > 0$, $m \geq 0$, $r(t)$ and $\rho(t)$ are positive continuous functions, and $I_0(t)$ is an admissible continuous nonnegative function, there exist unique continuous functions $S(t) > 0$, $I(t) > 0$, $\tau(t)$ which satisfy (3.1) (3.2) and (3.5). Further, these functions depend continuous on m, $\rho(t)$, $r(t)$, S_0, $I_0(t)$.

The content of Theorem 3.1 is that the above system constitutes a well-posed problem. Incidentally, by continuous dependence in the final statement of the theorem we mean to use the sup norm on a finite interval $[0,T]$, $(T > t_0)$. The proof will be omitted as in the next section we prove a more general theorem.

The special case that the ratio $\rho(x)/r(x)$ is constant leads to a considerable simplification. Let $K = \rho(x)/r(x)$. Since $S(t) > 0$, $t \geq 0$, equation (3.1) can be solved for $I(t)$, and this expression inserted in (3.5). It then follows that, for $t \geq t_0$,

$$m = \int_{\tau(t)}^{t} \rho(x)I(x)dx$$

$$= \int_{\tau(t)}^{t} - \frac{\rho(x)S'(x)}{r(x)S(x)} dx = -K \log \frac{S(t)}{S(\tau(t))}$$

or

$$S(\tau(t)) = e^{\nu} S(t)$$

where $\nu = m/K$. We note in passing that this is an example where ν of the preceding section has been explicitly determined. For $t \leq t_0$, $S(\tau(t)) = S_0$. Hence equation (3.1) then becomes

$$(3.6) \quad S'(t) = \begin{cases} -r(t)S(t)I_0(t), & 0 \leq t \leq t_0 \\ -r(t)S(t)[I_0(t) + S_0 - e^{\nu}S(t)], & t_0 \leq t \leq t_0 + \sigma \\ -r(t)S(t)e^{\nu}[S(t-\sigma) - S(t)], & t_0 + \sigma < t \end{cases}$$

$$S(0) = S_0.$$

This is a differential difference equation which could be solved by the method of steps once the initial function is given. On $[0,t_0]$, $S(t) = S_0 \exp -\int_0^t r(x)I_0(x)dx$. On $[t_0,\sigma]$, $S(t)$ is the solution of the Ricatti equation

$$S'(t) = -r(t)S(t)[I_0(t) + S_0 - e^{\nu}S(t)].$$

Thereafter solutions can be found successively on intervals $[j\sigma(j+1)\sigma]$, $j = 1,2,\cdots$, by the method of steps. (An analytic solution can also be found. See the comments at the end of the section.)

In Figures 3.1 - 3.3, some sample solutions are given to illustrate $S(t)$ and $I(t)$ for a variety of parameter values.

In this special case the analysis of the preceding section also applies if $r(t) \equiv$ constant, and we may compute the intensity. We already have $\nu = \frac{m}{K} = \frac{mr}{\rho}$. Further

$$E = \int_0^\infty P(x)dx = \int_0^\sigma 1 \cdot dx = \sigma.$$

Using (2.5) we need only to find the roots of

(3.7) $$F = \exp\{S_0 r\sigma(F - 1 - \epsilon)\}$$

where

$$\epsilon = \frac{\rho \int_0^\sigma I_0(x)dx - m}{S_0 \sigma \rho}$$

and apply (2.6) to find the intensity.

One can use the expression (3.7) to determine how the quantity F for this particular model changes with respect to the initial parameters r, m, ρ, and the initial "quantity" of infection $\int_0^\sigma I_0(x)dx$. For example, a straightforward calculation yields

$$\frac{\partial F}{\partial r} = \frac{FS_0\sigma(F-1-\epsilon)}{1 - FS_0 r\sigma}$$

$$= \frac{F}{r} \frac{\beta(F-1-\epsilon)}{1 - \beta F}.$$

Since by (2.6) $1 - \beta F > 0$, $F < 1$, and $\epsilon > 0$, $\frac{\partial F}{\partial r} < 0$. Similarly

$$\frac{\partial F}{\partial m} = \frac{rF}{\rho(1-\beta F)} > 0,$$

FIGURE 3.1

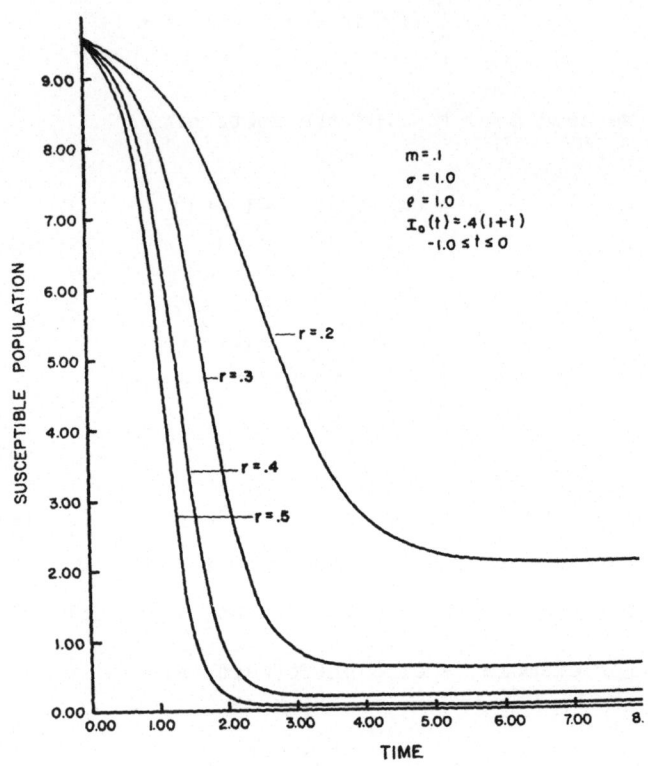

FIGURE 3.2

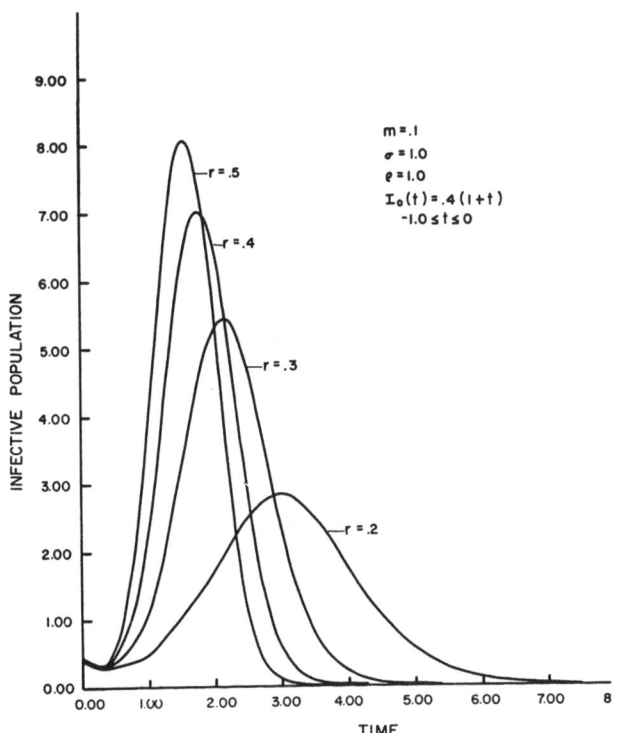

FIGURE 3.3

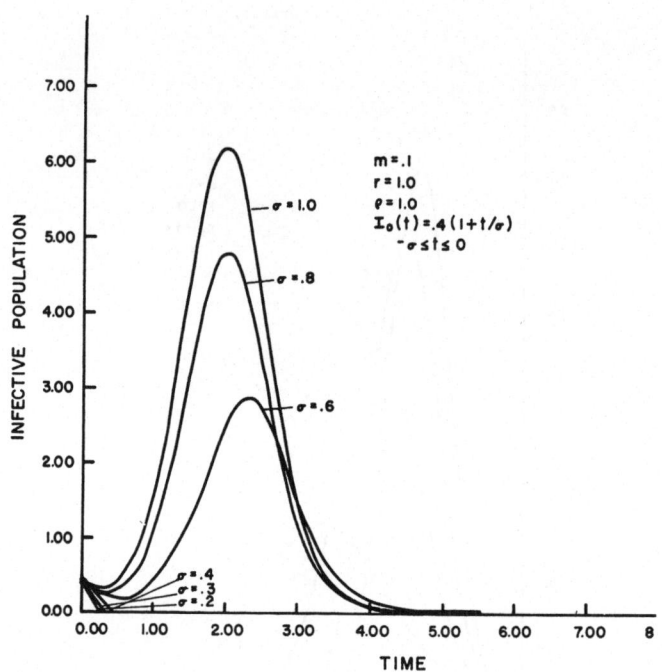

$$\frac{\partial F}{\partial \rho} = \frac{-mrF}{\rho^2(1-\beta F)} < 0,$$

and if $\gamma = \int_0^\sigma I_0(x)dx$

$$\frac{\partial F}{\partial \gamma} = \frac{-F}{S_0\sigma(1-\beta F)} < 0.$$

The signs of the above derivatives are all intuitively clear. The higher the contact rate, r, the virulence, ρ, or the initial "infective capacity," $\int_0^\sigma I_0(x)dx$, the larger the number who will be infected and hence the smaller F will be. On the other hand, with a higher threshold (higher natural resistance), the fewer the number expected to be infected.

If $\rho(t), r(t)$ are arbitrary positive functions, the above procedure for finding the intensity does not apply. However, since there does exist a positive solution $I(t)$, it follows that $\tau(t)$ is monotone increasing and thus $\lim_{t\to\infty} \tau(t)$ exists or diverges to $+\infty$. Thus if $S(\infty) \neq 0$, there exists a ν such that $S(\tau(\infty)) = e^\nu S(\infty)$. While ν is unknown, bounds can sometimes be found.

Suppose, for example, that r is constant and that $0 < K_1 \leq \rho(t) \leq K_2$. As noted above, the primary interest in $\rho(t)$ and $r(t)$ was to allow for periodic variations, so if $\rho(t)$ was periodic then K_1 and K_2 above would exist. Then proceeding as before one obtains the inequalities

$$e^{\nu_2} S(t) \leq S(\tau(t)) \leq e^{\nu_1} S(t)$$

where $\nu_1 = \frac{mr}{K_1}$ and $\nu_2 = \frac{mr}{K_2}$.

Denote the true value of ν by ν_3 and let

$$f_i(z) = z - \exp(\beta(z - 1 - \epsilon_i)), \quad i = 1, 2, 3$$

where

$$\epsilon_i = \frac{r \int_0^\sigma I_0(x) dx - \nu_i}{\beta} \, .$$

Let z_i denote the zero of f_i. Then

$$0 = f_1(z_1) \geq f_3(z_1),$$

and

$$0 = f_2(z_2) \leq f_3(z_2),$$

so

$$z_1 \leq z_3 \leq z_2 \, .$$

Thus since ν_1 and ν_2 are known, Figure 2.1 could be used to find bounds for the value of F and hence for the intensity of the epidemic.

Notes:

A threshold of the form (iii) for epidemic models was proposed by

K. L. Cooke, Functional differential equation: Some models and perturbation problems, in Differential Equations and Dynamical Systems (J. K. Hale and J. P. LaSalle, eds.), Academic Press, New York, 1967.

The model there had the initial infectives remaining in the population as infectives for all time. The question of existence of solutions was posed as an open problem.

The model in this section, with recovery of initial infectives, appeared in

> F. Hoppensteadt and P. Waltman, A Problem in the
> Theory of Epidemics, Math. Biosciences 9(1970), 71-91.

The figures are from this paper. Existence, uniqueness, and continuous dependence of solutions was established.

The limiting value of the infective population when r and ρ are constant was first determined by

> H. Hethcote, Note on Determining the Limiting Susceptible Population in an Epidemic Model, Math.
> Biosciences 9(1970), 161-163.

A complete analysis of this special case of the model was presented in

> L. O. Wilson, An Epidemic Model Involving a Threshold, Math. Biosciences 15(1972), 109-121,

where an exact solution was found and an asymptotic estimate provided of the approach to the limit, and the changes with respect to parameters investigated.

As noted above, it is desirable to have $\rho(x)$ and $r(x)$ functions and not constants so that one might take into account seasonal variations. Although the existence theory covers this case, any determination of asymptotic behavior remains an open question.

The "method of steps" is described in

> L. E. El'sgol'ts, Introduction to the theory of Differential Equations with Deviating Arguments,
> Holden-Day, Inc., 1966.

4. A THRESHOLD MODEL WITH TEMPORARY IMMUNITY

In the previous models it was assumed that contracting the disease provided permanent immunity and hence the class R of removed individuals was the final resting place for all individuals who became infective. With many diseases, however, the individual, after recovering from an infection, may become susceptible again. We now consider a model, described symbolically,

$$S \to E \to I \to R \to S,$$

in which the individual in R is temporarily immune to the infection but eventually returns to the susceptible class. Although the description of the model differs from the previous one only on this point, the resulting model produces significantly more difficult mathematical problems.

We now assume the four classes (S),(E),(I),(R) as before where now (R) is recovered and immune. The infection spreads according to the following rules:

(i) The rate of exposure of susceptibles to infectives at time t is given by $-r(t)S(t)I(t)$.

(ii) An individual who is first exposed at time τ becomes infective at time t if

$$\int_{\tau}^{t} [\rho_1(x) + \rho_2(x)I(x)]dx = m$$

where $\rho_1(x)$, $\rho_2(x)$ are given nonnegative functions and m is a nonnegative constant.

(iii) An individual who becomes infective at time t
       recovers from the infection at time t + σ, σ
       a given positive constant.

(iv) An individual who becomes infective at time t
      is immune until t + ω, at which time he becomes
      susceptible again, where ω is a given nonnega-
      tive constant.

(v) The population size is constant.

In (ii) a more general threshold has been incorporated to allow
for an incubation period. As noted in the preceding section,
$\rho_1(t)$ = constant, $\rho_2(t) \equiv 0$, yields $t - \tau(t)$ constant.

Hypothesis (iv) states that the length of time spent in class
(R) is precisely ω. The model of Section 3 results from formally
setting ω = +∞. The other hypotheses are exactly as in the previous
case.

As before, we assume as initial conditions S_0, the number of
susceptibles, and $I_0(t)$, $-\sigma \leq t \leq 0$, the past history of the infectives which are inserted into the susceptible population at $t = 0$.
Assuming that (iii) applies to these initial infectives as well, the
future of this group is again known. We take this into account by
defining an extension of $I_0(t)$ to the real line by

$$I_0(t) = \begin{cases} 0, & |t| \geq \sigma \\ I_0(t), & -\sigma \leq t \leq 0 \\ I_0(0) - I_0(t-\sigma), & 0 \leq t \leq \sigma. \end{cases}$$

We also assume the existence of $t_0 < \sigma$ such that

$$\int_{0.}^{t_0} [\rho_1(x) + \rho_2(t)I_0(x)]dx = m.$$

A monotone function, $I_0(t)$, $-\sigma \leq t \leq 0$, $I_0(-\sigma) = 0$, whose extension satisfies the above integral condition will be said to be <u>admissible</u>. This is the threshold condition.

The equations for the model will be derived from the rate at which individuals are leaving the susceptible class. It is necessary first of all to account for those initially infective individuals who are in classes (S) and (R) for the first time. These will be denoted by $I_1(t)$ and $I_2(t)$, respectively. The earliest possible entry for an initial infective into class (S) is ω, and for $t > \omega$, the number of initially infective individuals who are in (S) for the first time is the number who recover before $t - \omega$. Thus

$$I_1(t) = \begin{cases} 0, & t \leq \omega \\ I_0(0) - I_0(t-\omega), & \omega \leq t. \end{cases}$$

Similar arguments for $I_2(t)$ yield

$$I_2(t) = \begin{cases} I_0(0) - I_0(t), & t \leq \omega \\ I_0(t-\omega) - I_0(t), & \omega \leq t. \end{cases}$$

The function $\tau(t)$ must satisfy

(4.1)
$$\int_{\tau(t)}^{t} [\rho_1(x) + \rho_2(x)I(x)]dx = m,$$

$$\tau(t) = 0, \quad t \leq t_0.$$

The second equation for $\tau(t)$ is just a convenience for eliminating special cases in the remaining equations. The susceptible population at time t is everyone who has not been exposed in the interval $(\tau(t - \sigma - w),t)$. Certainly exposure in that interval eliminates an individual from (S) and exposure before $\tau(t - \sigma - w)$ results in infection before time $t - \sigma - w$, recovery before $t - w$, and readmission to (S) before t. Thus it follows that

$$(4.2) \qquad S(t) = I_1(t) + S_0 - \int_{\tau(t-\sigma-w)}^{t} (rIS)(x)dx,$$

where $rIS(x) = r(x)I(x)S(x)$. Note that if $w = +\infty$, this is

$$S(t) = S_0 - \int_0^t (rIS)(x)dx$$

or

$$S'(t) = -r(t)I(t)S(t),$$

the equation for the rate of change of susceptible individuals in the model of Section 3.

In the same way, the number of infectives at time t consists of the initial infectives who are still infective at time t plus those individuals who were exposed between $\tau(t-\sigma)$ and $\tau(t)$. Those first exposed before $\tau(t-\sigma)$ have already recovered and those first exposed after $\tau(t)$ are not yet infective. This is expressed by

$$(4.3) \qquad I(t) = I_0(t) + \int_{\tau(t-\sigma)}^{\tau(t)} (rIS)(x)dx.$$

The number of individuals in the exposed class at time t are those who were first exposed after $\tau(t)$, or

$$(4.4) \qquad\qquad E(t) = \int_{\tau(t)}^{t} (rIS)(x)dx.$$

Finally, since the population is constant, it follows that

$$(4.5) \qquad\qquad R(t) = N - S(t) - E(t) - I(t)$$

$$= I_2(t) + \int_{\tau(t-\sigma-\omega)}^{\tau(t-\sigma)} (rIS)(x)dx.$$

Equations (4.1) - (4.5) constitute the model. As before, it is sufficient to establish the existence of $\tau(t), S(t), I(t)$ for equations
(4.4) (4.5) yield $E(t)$ and $R(t)$ directly in terms of these quantities by an integration.

The first mathematical question of interest is whether or not
there exists a solution to such equations, and if so, whether it is
unique. One would also like to know something about the ultimate
(limiting) behavior of the solutions, in particular, what sort of
recurrence properties can be established. It is also of interest to
find the solutions numerically. In this section we consider only
the existence and uniqueness question — most of the rest remains
open (see Section 5).

THEOREM 4.1. Let $I_0(t)$ be admissible with $I_0(t_0) > 0$, and let
$r > 0$, $\rho_1 \geq 0$, $\rho_2 > 0$ be continuous functions and let $\sigma > 0$, $\omega \geq 0$,
$m \geq 0$ be given constants. Then there exist unique continuous func-
tions τ, S, E, I, R which satisfy (4.1) - (4.5) on $[0, \infty)$. Further,
$I(t)$ and $S(t)$ are positive. Solutions depend continuously on the
initial conditions $I_0(t)$, S_0 and on the choice of r, ρ_1, ρ_2, σ,
ω, and m.

Proof: The basic idea of the proof is to show that the system
(4.1) - (4.3) has a solution on a small closed interval where the
problem is easy, and then to show that if a solution exists on an
arbitrary closed interval, it can be extended to a larger closed interval. The proof is completed by establishing that if the solution
exists on some maximal (open) interval it can be extended to a closed
interval.

On the interval $[0,t_0]$ no new infectives occur, that is,
$\tau(t) \equiv 0$ and $I(t) = I_0(t) > 0$. In this case $S(t)$ can be found as
the solution of

$$S(t) = I_1(t) + S_0 - \int_0^t r(x)I_0(x)S(x)dx,$$

a __linear__ integral equation. (If $t_0 < \omega$, this would be an ordinary
differential equation.)

Suppose now that a solution $S(t), I(t), \tau(t)$ has been found on
$[0,t_1]$, with $S(t) > 0$, $I(t) > 0$ (and hence $\tau(t)$ nondecreasing,
in fact increasing if $t > t_0$). We wish to extend this solution to
$[0,t_2]$, $t_2 > t_1$, where t_2 will be selected below. Denote the set
of continuous functions on $[t_1,t_2]$ with uniform norm by $C[t_1,t_2]$
where we require a priori, $t_2 - t_1 < \sigma$ (other restrictions will be
imposed on the size of $t_2 - t_1$ in the course of the proof). Let

$$\mathfrak{m} = [\varphi(t) \,|\, \varphi \in C[t_1,t_2],\ \gamma \le \varphi(t) \le N,\ \varphi(t_1) = I(t_1)]$$

where $\gamma = 1/2\, I_0(t_0)$ if $t_1 = t_0$ or $\gamma = r_* S_* I_* [\tau(t_1) - \tau(t_2 - \sigma)]$ if
$t_1 > t_0$, and $f_*(f^*)$ is used to denote the minimum (maximum) of a
continuous function on a given closed interval — in this case,
$[0,t_1]$. We note that γ is strictly positive. In case $t_1 = t_0$,
$I_0(t_0) > 0$ by hypothesis; if $t_1 > t_0$, since $\tau(t)$ on $(t_0,t_1]$

satisfies

$$\tau'(t) = \frac{\rho_1(t) + \rho_2(t)I(t)}{\rho_2(\tau) + \rho_2(\tau)I(\tau)} > 0,$$

and $t_2 - \sigma < t_1$, it follows that $\tau(t_1) - \tau(t_2-\sigma) > 0$. Equip $\mathcal{M}$ with the metric $\rho(x,y) = \max_{[t_1,t_2]} |x(t) - y(t)|$, i.e., $\rho(x,y) = \|x - y\|$ where $\|\cdot\|$ is norm of $C[t_1,t_2]$. $\mathcal{M}$ is a complete metric space since convergence in this norm is uniform convergence of functions. The mapping we wish to define on $\mathcal{M}$ involves an intermediate mapping. Let $t_2 - t_1 < \dfrac{m}{\rho_1^* + \rho_2^* N}$ and on $\mathcal{M}$, define a mapping $U(\varphi)$ by

$$\int_{U\varphi(t)}^{t_1} [\rho_1(x) + \rho_2(x)I(x)dx] + \int_{t_1}^{t} [\rho_1(x) + \rho(x)\varphi(x)]dx = m.$$

Since $\rho_2(x)\varphi(x) > 0$, and $t > t_0$, the restriction on $t_2 - t_1$ will make $(U\varphi)(t)$ uniquely defined. Also, it is the case that $U\varphi(t_1) = \tau(t_1)$.

LEMMA 4.1. If $t_2 - t_1 < \dfrac{m}{\rho_1^* + \rho_2^* N}$, U is a continuous mapping of $\mathcal{M}$ into $C[t_1,t_2]$ with $\tau(t_1) \le U\varphi(t) \le t_1$.

Proof: $U\varphi(t) > \tau(t_1)$ since $U\varphi(t_1) = \tau(t_1)$ and $\rho_2(t)\varphi(t) > 0$. From the definition of U, it follows that

$$m = \int_{U\varphi(t)}^{t_1} [\rho_1(x) + \rho_2(x)I(x)]dx + \int_{t_1}^{t} [\rho_1(x) + \rho_2(x)\varphi(x)]dx$$

$$\le (\rho_1^* + \rho_2^* N)(t_2 - U\varphi(t)),$$

or

$$U\varphi(t) \leq t_2 - \frac{m}{\rho_1^* + \rho_2^* N} \leq t_1 .$$

Let $\varphi_1, \varphi_2 \in \mathcal{M}$. Then,

$$\int_{U\varphi_1(t)}^{t_1} [\rho_1(x) + \rho_2(x)I(x)dx] + \int_{t_1}^{t} [\rho_1(x) + \rho_2(x)\varphi_1(x)]dx$$

$$= \int_{U\varphi_2(t)}^{t_1} [\rho_1(x) + \rho_2(x)I(x)]dx + \int_{t_1}^{t} [\rho_1(x) + \rho_2(x)\varphi_2(x)]dx.$$

Rearranging the integrals yields

$$\int_{t_1}^{t} \rho_2(x)(\varphi_1(x) - \varphi_2(x))dx = \int_{U\varphi_2(t)}^{U\varphi_1(t)} [\rho_1(x) + \rho_2(x)I(x)]dx,$$

or

$$\rho_2^* \rho(\varphi_1, \varphi_2)|t_2 - t_1| \geq \rho_{2*} I_* |U\varphi_1(t) - U\varphi_2(t)|.$$

Taking the maximum of both sides on $[t_1, t_2]$ gives

(4.6) $$\|U\varphi_1 - U\varphi_2\| \leq K |t_2 - t_1| \rho(\varphi_1, \varphi_2)$$

where $K = \rho_2^* / \rho_{2*} I_*$. Hence U is continuous.

Define $T : \mathcal{M} \to C[t_1, t_2]$ by

$$(T\varphi)(t) = I_0(t) + \int_{\tau(t-\sigma)}^{U\varphi(t)} (rIS)(x)dx.$$

LEMMA 4.2. <u>For</u> <u>sufficiently</u> <u>small</u> $t_2 - t_1$, T <u>defines</u> <u>a</u> <u>contraction</u> <u>mapping</u> <u>of</u> $\mathcal{M}$ <u>into</u> <u>itself</u>.

Proof: T is a continuous function U and τ are continuous. Further

$$(T\varphi)(t) \geq I_0(t) + r_* I_* S_* [U\varphi(t) - \tau(t-\sigma)]$$

$$\geq I_0(t) + r_* I_* S_* [\tau(t_1) - \tau(t_2-\sigma)]$$

$$\geq \gamma$$

provided

$$t_2 - t_1 \leq \min\left(\frac{m}{\rho_1^* + \rho_2^* N}, \sigma\right)$$

or if $t_1 = t_0$, t_2 is so small that $I_0(t_2) > I_0(t_0)/2$. On the other hand,

$$(T\varphi)(t) = I_0(t) + \int_{\tau(t-\sigma)}^{U\varphi(t)} (rIS)(x)dx$$

$$\leq I_0(t) + \int_{\tau(U\varphi(t)-\sigma-\omega)}^{U\varphi(t)} (rIS)(x)dx$$

since $(U\varphi)(t) < t$, $\tau(t)$ is monotone, and $\omega \geq 0$. Thus it follows that

$$T\varphi(t) \leq I_0(t) + S_0 + I_1[U\varphi(t)] - S[U\varphi(t)]$$

$$\leq I_0(t) + S_0 + I_1[U\varphi(t)]$$

since the integral on the right hand side of the inequality is the integral on the right hand side of (4.2) and $U\varphi(t) < t_1$. Further

$$I_0(t) + S_0 + I_1[U\varphi(t)] = \begin{cases} I_0(t) + S_0, & \text{if } U\varphi(t) < \omega \\ \\ I_0(t) + S_0 + I(0) - I[U\varphi(t) - \omega], & \text{if } U\varphi(t) \geq \omega. \end{cases}$$

Since for $U\varphi(t) \geq \omega$

$$I_0(t) - I_0[U\varphi(t) - \omega] \leq 0,$$

it follows in either case that

$$T\varphi(t) \leq N.$$

Since $(T\varphi)(t_1) = I(t_1)$, it has been shown that

$$T\mathcal{M} \subset \mathcal{M}.$$

To show that the mapping T is a contraction for sufficiently small $t_2 - t_1$, we note that for $\varphi_1, \varphi_2 \in \mathcal{M}$,

$$|T\varphi_1(t) - T\varphi_2(t)| = \left| \int_{U\varphi_1(t)}^{U\varphi_2(t)} (rIS)(x)dx \right|$$

$$\leq r^*N^2 |U\varphi_1(t) - U\varphi_2(t)|$$

$$\leq r^*N^2 \|U\varphi_1 - U\varphi_2\|.$$

Taking the maximum on $[t_1, t_2]$ gives

$$\rho(T\varphi_1, T\varphi_2) \leq r^*N^2 \|U\varphi_1 - U\varphi_2\|$$

$$\leq r^*N^2 K |t_2 - t_1| \rho(\varphi_1, \varphi_2)$$

where (4.6) has been used to obtain the second inequality. If

$$|t_1 - t_2| < \min\left[\sigma, \frac{m}{\rho_1^* + \rho_2^* N}, \frac{1}{r^* K N^2}\right]$$

then all of the above requirements on the size of $t_2 - t_1$ are met and T is a contraction. (If $t_1 = t_0$, then t_2 so small that $I_0(t_2) > I_0(t_0)/2$ needs to be added to the above set of conditions.)

By the contraction mapping theorem there exists a unique point φ such that $T\varphi = \varphi$. Extend the given functions $\tau(t)$, $I(t)$, known on $[0,t_1]$ to $[0,t_2]$ by defining them on $[t_1,t_2]$ as

$$I(t) = \varphi(t)$$

$$\tau(t) = U\varphi(t).$$

To find $S(t)$ it is necessary only to solve the linear integral equation (linear since $\tau(t)$ and $I(t)$ are now known functions)

$$(4.7) \qquad S(t) = I_1(t) + S_0 - \int_{\tau(t-w-\sigma)}^{t} rIS(x)dx.$$

Assuming that a solution to (4.7) exists, $E(t)$ and $R(t)$ can be found by evaluating the right hand side of (4.4) and (4.5). Subject to the above reservation, the existence of a solution has been established on $[0,t_2]$.

The above extension process can be continued to a larger interval, and then to a larger interval, etc. If the process does not cover the ray $[0,\infty)$, then the right hand endpoints of the extension intervals have an accumulation point and hence there is a maximal right open interval $[0,T)$ to which a solution can be extended. Since $\tau(t)$ is monotone increasing, $\tau(T^-)$ exists. Further, since

$$I(t) = I_0(t) + \int_{\tau(t-\sigma)}^{\tau(t)} rIS(x)dx,$$

then

$$I(T^-) = I_0(T) + \int_{\tau(T-\sigma)}^{\tau(T^-)} rIS(x)dx.$$

Since the integrand is positive and $\tau(T^-) > \tau(T-\sigma)$ (recall that $\sigma > 0$), then $I(T^-) > 0$. In the same way since $|S(t)| \leq N$, we can define $S(T^-)$ by

$$S(T^-) = I_1(t) + S_0 - \int_{\tau(T-\sigma-w)}^{T^-} rIS(x)dx.$$

Thus S, I, and τ can be defined as continuous functions on $[0,T]$.

That $I(T) > 0$ was immediate but it is not clear that $S(T) > 0$. We use a differential inequality to establish this fact. A small technical difficulty occurs in that $I_1(t)$ may not be differentiable. However, it is monotone increasing and hence differentiable except for a set of measure zero (a.e.). Thus

$$S'(t) \geq -rIS(t) + rIS(\tau(t-\sigma-w)), \text{ a.e.}$$

$$\geq -r(t)I(t)S(t), \text{ a.e.},$$

or

$$S(T) \geq S_0 \exp\left(-\int_0^T r(x)I(x)dx\right) > 0.$$

(The a.e. can be dropped in the last step since $S(t)$ is continuous.) Thus existence can be extended to $[0,T]$ with $I(T) > 0$,

and hence can be extended to an even larger interval, contradicting
the maximality of T.

On the initial interval $[0,t_0]$, the solution was unique and
each extension produced a unique solution, hence the uniqueness asser-
tion of the theorem.

There remains to be shown that (4.7) has a solution on $[0,t_2]$
given $\tau(t), I(t)$ on $[0,t_2]$ and $S(t)$ on $[0,t_1]$. For $t \in [0, \sigma + w]$,
(4.7) is a linear Volterra integral equation

$$(4.8) \qquad S(t) = I_1(t) + S_0 - \int_0^t (tIS)(x)dx,$$

since $I(x)$ is known. On $[\sigma + w, \infty)$, $I_1(t) \equiv I_0$, so that (4.7) can
be written

$$S(t) = N - \int_0^t rIS(x)dx + \int_0^{\tau(t-w-\sigma)} rIS(x)dx$$

or

$$(4.9) \qquad S(t) = N - \int_0^t rIS(x)dx + f(t),$$

where $f(t)$ is a known function since $\tau(t) \le t_1$ for $t_1 \le t \le t_2$,
and $S(t)$ is known on $[0,t_1]$. This again is a linear Volterra
integral equation. This settles the existence and uniqueness question.

The continuous dependence arguments are tedious (but straightforward) and are omitted.

The requirement $\rho_2(t) > 0$ does not allow for a constant incubation period. The following theorem exchanges the condition $\rho_2(t) > 0$
for $\rho_1(t) > 0$.

THEOREM 4.2. Let $I_0(t)$ be an admissible function and let $r > 0$, $\rho_1 > 0$, $\rho_2 \geq 0$ be continuous functions and let $\sigma > 0$, $m \geq 0$, $\omega \geq 0$ be given constants. Then there exist unique functions τ, S, E, I, R which satisfy (4.1) - (4.5) on $[0, \infty)$. Solutions depend continuously on the initial conditions $I_0(t), S_0$ and on the choice of $r, \rho_1, \rho_2, \sigma, \omega,$ and m.

The proof of this theorem is omitted. It is similar to (and easier than) the proof given. Moreover, in Section 6, a proof will be given with $\rho_1 > 0$, $\rho_2 \geq 0$ for a two population model which exhibits the necessary ideas.

Notes:

The model and theorems in this section can be found in

F. Hoppensteadt and P. Waltman, A Problem in the Theory of Epidemics II, Math. Biosciences 12(1971), 133-145.

5. SOME SPECIAL CASES AND SOME NUMERICAL EXAMPLES

In the preceding section a model was proposed which essentially involved the following three equations in $\tau(t)$, $I(t)$ and $S(t)$ for large t,

$$(5.1) \qquad \int_{\tau(t)}^{t} [\rho_1(x) + \rho_2(x)I(x)]dx = m,$$

$$\tau(t) \equiv 0, \qquad 0 \leq t \leq t_0 < \sigma,$$

$$(5.2) \qquad S(t) = I_1(t) + S_0 - \int_{\tau(t-\sigma-\omega)}^{t} r(x)S(x)I(x)dx,$$

$$(5.3) \qquad I(t) = I_0(t) + \int_{\tau(t-\sigma)}^{\tau(t)} r(x)S(x)I(x)dx.$$

($I_0(t)$ and $I_1(t)$ were described in the previous section.) The questions of the existence, uniqueness, and continuous dependence of solutions were resolved so the model is mathematically sensible, but many other questions of interest in studying epidemics were left open.

Foremost among these are questions involving limiting behavior and the development of numerical techniques for computing solutions. Since the supply of susceptibles is replenished from the removed class, some sort of recurrence is not unexpected. Such recurrence is observed for example when a disease infects large numbers of people, is dormant for a number of years, and then reappears. Such can be explained by temporary immunity which inhibits the occurrence of an epidemic, but as the immunity is lost, the potential for a large out-break reappears. Such would seem to fall within the scope of the

model being considered. As far as the author is aware, no progress
has been made on the determination of limiting behavior for these
equations. In particular it is not known whether (5.1) (5.2) (5.3)
has a solution with I(t) periodic (or with any other specific recurrence property). We will return to this particular question at the
end of this section with some numerical evidence.

We look first at some special cases of the model in order to
relate it to other known work. Suppose that $m = 0$ (no threshold),
and $\omega = 0$ (instant recovery). Without the threshold $\tau(t) \equiv t$ and
class E has no members while $\omega = 0$ makes class R empty as well.
Schematically this is represented by

$$S \rightarrow I \rightarrow S.$$

The equations become (using the definition of $I_1(t)$)

$$S(t) = I_0 - I_0(t) + S_0 - \int_{t-\sigma}^{t} rIS(x)H(x)dx,$$

$$I(t) = I_0(t) + \int_{t-\sigma}^{t} rIS(x)H(x)dx,$$

$$E(t) = R(t) = 0,$$

where $H(x) = 0$, $x \geqslant 0$ and $H(x) = 1$, $x < 0$. Since $S(t) + I(t) = N$,
there is really only one equation, valid for all t,

(5.4) $$I(t) = I_0(t) + \int_{t-\sigma}^{t} H(x)r(x)I(x)[N - I(x)]dx.$$

The basic assumption on the length of the infectious period requires
$I_0(t) \equiv 0$ $t > \sigma$, and this equation becomes

$$(5.5) \qquad I(t) = \int_{t-\sigma}^{t} r(x)I(x)[N - I(x)]dx, \qquad t > \sigma.$$

Equations like (5.5) have been suggested as a model for gonorrhea where infectivity occurs with contact and where there is negligible immunity. In particular, Cooke and Yorke have studied the equations

$$(5.6) \qquad x(t) = c + \int_{t-\sigma}^{t} g(x(s))ds$$

and

$$(5.7) \qquad x(t) = c + \int_{t-\sigma}^{t} P(t-s)g(x(s))ds, \qquad t \geq t_0.$$

Their principal result is:

THEOREM 5.1. <u>Assume that</u> $g(x)$ <u>is a continuously differentiable function and that</u> $P(t)$ <u>is continuously differentiable, nonincreasing, and nonnegative on</u> $0 \leq t \leq \sigma$. <u>Let</u> $x(t)$ <u>be any solution of</u>

$$x(t) = c + \int_{t-\sigma}^{t} P(t-s)g(x(s))ds, \qquad t > \eta,$$

<u>and let</u> $[\eta - \sigma, T)$ <u>be its maximal interval of existence, where</u> $\eta < T \leq \infty$. <u>Then one of the following holds</u>:

    (i) $x(t) \to \infty$ <u>as</u> $t \to T$,

    (ii) $x(t) \to$ constant <u>as</u> $t \to T$,

    (iii) $x(t) \to -\infty$ <u>as</u> $t \to T$.

In the model in the epidemic case, alternatives (i) and (iii) cannot occur, and since $T = \infty$, (ii) shows that solutions tend asymptotically to a limiting constant as $t \to \infty$. Cooke has further studied the asymptotic behavior of the equations

(5.8)
$$x(t) = \int_{t-\sigma}^{t} P(t-s)g(x(s))ds + f(t)$$

and

(5.9)
$$x(t) = \int_{t-\sigma}^{t} [P(t-s)g(x(s)) + f(s)]ds.$$

The forcing term $f(t)$ can be thought of, in the epidemic model, as immigrations. Cooke's results make use of more sophisticated mathematical tools and so are somewhat technical to state; the reader is referred to the original paper (see Notes for reference) for more details. Both sets of equations also have population growth and economic interpretations.

Another special case of interest is $\rho_1 \equiv$ constant, $\rho_2 \equiv 0$, $\sigma = +\infty$. The model is schematically represented

$$S \rightarrow E \rightarrow I.$$

It follows immediately that $\tau(t) = t - m/\rho_1$ and hence since $I_1(t) \equiv 0$ (no initial infectives can become susceptible),

$$S'(t) = -r(t)S(t)I(t),$$

$$I(t) = I_0(t) + \int_0^{t-m/\rho_1} rIS(x)dx$$

$$= I_0(0) - S(t-m/\rho_1) + S(0)$$

$$= N - S(t-m/\rho_1), \quad t > m/\rho_1.$$

The initially infective individuals remain infective, hence

$I_0(t) \equiv I_0(0) = N - S_0$ was used above. Replacing $I(t)$ in the equation for $S'(t)$ gives a difference-differential equation

$$(5.10) \qquad S'(t) = -r(t)S(t)[N - S(t-m/\rho_1)], \quad \text{for} \quad t > m/\rho_1.$$

The initial condition for this equation is the solution of the initial value problem

$$S' = -rIS(0)$$

$$S(0) = S_0$$

on $[0, m/\rho_1]$, or

$$S(t) = S_0 \exp\left(-I_0 \int_0^t r(x)dx\right).$$

Equation (5.10) is a type of delayed logistics equation and has been studied in many contexts other than epidemiological ones (see Notes).

The numerical solution of functional differential equations is only beginning to be studied (for sufficiently large t, i.e., after $I_0(t), I_1(t)$ have become constant (5.1) - (5.3) can be differentiated). Since the proof of existence was by a contraction mapping argument, construction of a solution by iterating the mapping is theoretically possible. It is "practically possible" as well if one does not ask for too long a time interval and has sufficient patience (and a sufficient computer budget). Direct schemes however are more attractive.

For $t < \sigma$, numerical solution is a straightforward problem since there are no delays. Assume r is constant and the initial

$I_0(t)$ is differentiable on $-\sigma \le t \le 0$. The differentiated form of the equation is given by (taking $\rho_1 \equiv 0$, $\rho_2 \equiv 1$ for simplification)

$$\tau'(t) = I(t) / I(\tau(t))$$

$$S'(t) = I_1'(t) - r[I(t)S(t) - I(\tau(t-\sigma-w))(\tau(t-\sigma-w))\tau'(t-\sigma-w)]$$

$$= I_1'(t) - r[I(t)S(t) - S(\tau(t-\sigma-w)I(t-\sigma-w)]$$

$$I'(t) = I_0'(t) + r[\ I(\tau(t))S(\tau(t))\tau'(t) - I(\tau(t-\sigma))S(\tau(t-\sigma))\tau'(t-\sigma)].$$

$$\doteq I_0'(t) + r[I(t)S(\tau(t)) - I(t-\sigma)S(\tau(t-\sigma))].$$

(At $t=0$ and $t=\sigma$, $I_0(t)$ is not differentiable, hence at $t=w$ and $t=\sigma+w$, $I_1(t)$ is not differentiable. At the first point we take the left hand derivative and at the second, the right hand derivative.) J. Mosevich has proposed using the right hand side to compute the derivative and then using the midpoint method, i.e.,

$$f(t) = f(t-2h) + 2hf'(t-h)$$

as predictor for S, I and τ. The integral form with a quadrature formula would be used as a corrector. In particular this predictor gives a way of advancing $\tau(t)$ — the advanced value being needed in the limits of integration for the integral equation form. The predicted values of $S(t), I(t), \tau(t)$ can be used, with the already known past history to compute the integrals on the right hand side of (5.2)(5.3) directly to obtain corrected values of $S(t), I(t)$. Finally, a new $\tau(t)$ can be obtained by solving

$$\int_x^t I(s)ds - m = 0$$

by a root finding process. (Regula Falsi was actually used.) The
corrector could be iterated if necessary for error control.

The procedure seems to work quite well. Some of the results are
presented graphically in Figures 5.1 - 5.4. The upper curves are the
susceptible population and the lower curves, the infected population.

In Figure 5.1, $\omega = 0$, $\sigma = 1$, $m = 0$. This is the case covered by
Theorem 5.1 and the convergence to a limiting value is quite rapid.
In Figures 5.2 - 5.4, $\sigma = 1.0$ and $m = .1$ are fixed but ω is varied.
For $\omega = 1$, oscillations die out rapidly (Figure 5.3). For $\omega = 2$,
the oscillations clearly are damped (Figure 5.3). For $\omega = 3$, the
solution looks suspiciously periodic (Figure 5.4). The solution has
been carried farther than illustrated and still has the same appear-
ance, although, of course, no matter how far one computes, damping
could begin on the next segment. (Moreover, the propagation of
errors in the numerical scheme is not known.) However, the graphs
strongly suggest the existence of a constant limit in some cases and
the possibility that periodic solutions may occur for critical values
of the parameter ω. This provides an interesting and open mathematical problem.

Notes:

The two special cases (5.4) and (5.10) were discussed in

> F. Hoppensteadt and P. Waltman, A System of Inte-
> gral Equations Describing a Deterministic Epi-
> demic Model, Lecture Series No. 15, Institute for
> Fluid Dynamics and Applied Mathematics, Univ. of
> Maryland, 1971, 25-28.

FIGURE 5.1

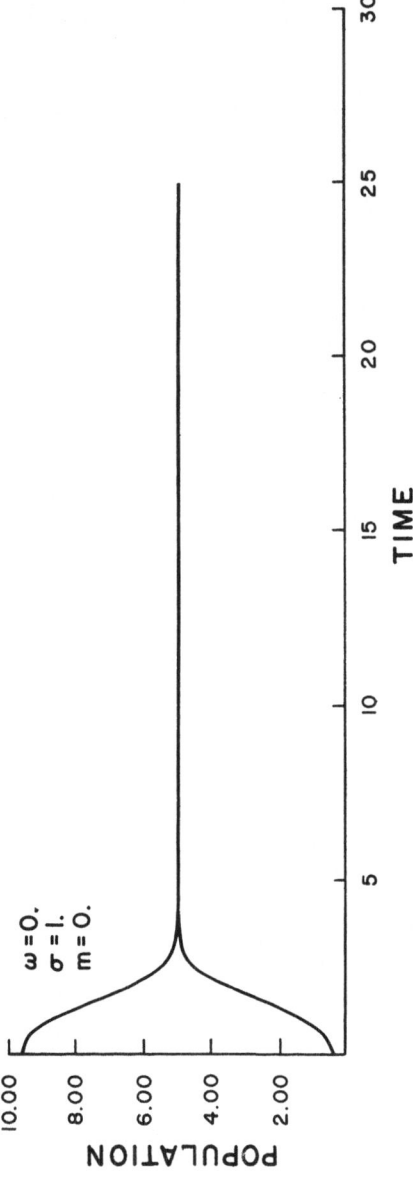

POPULATION

TIME

w=0.
b=1.
m=0.

FIGURE 5.2

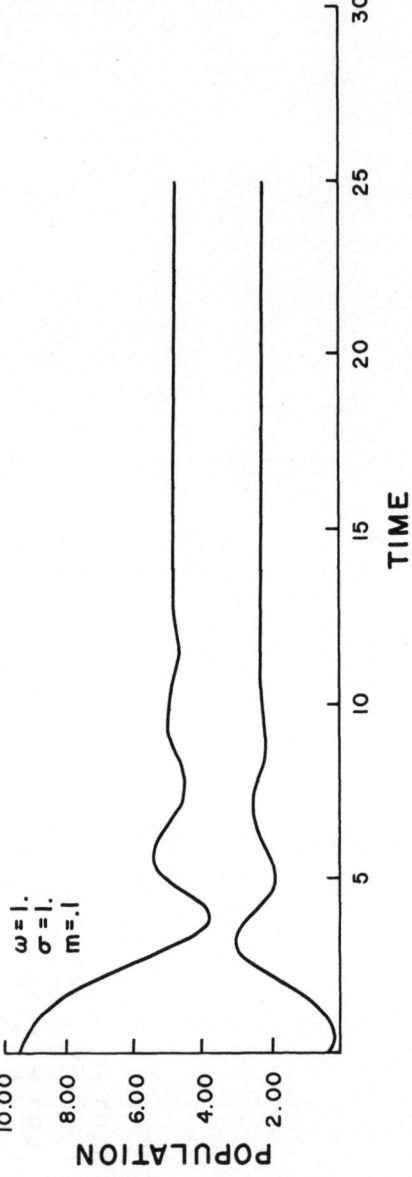

FIGURE 5.3

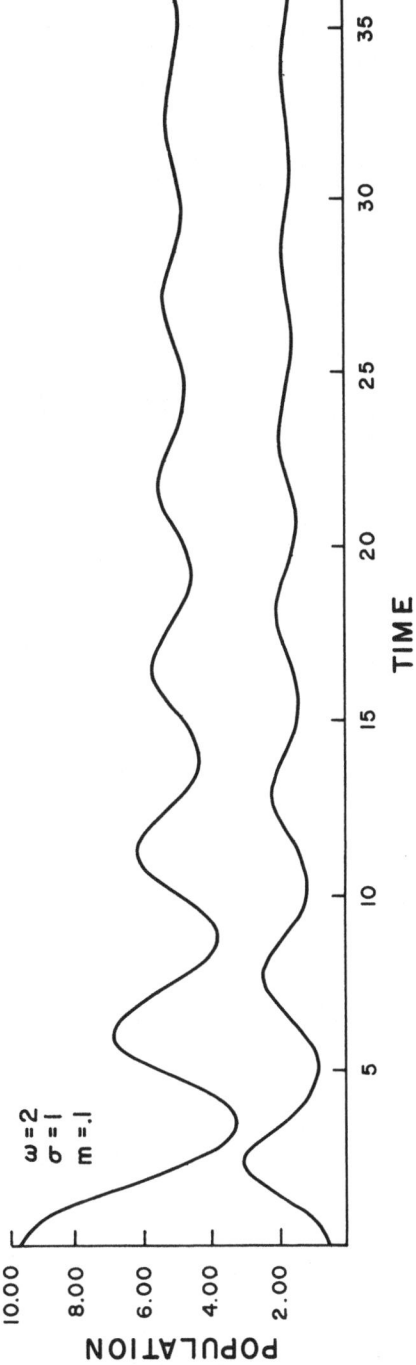

FIGURE 5.4

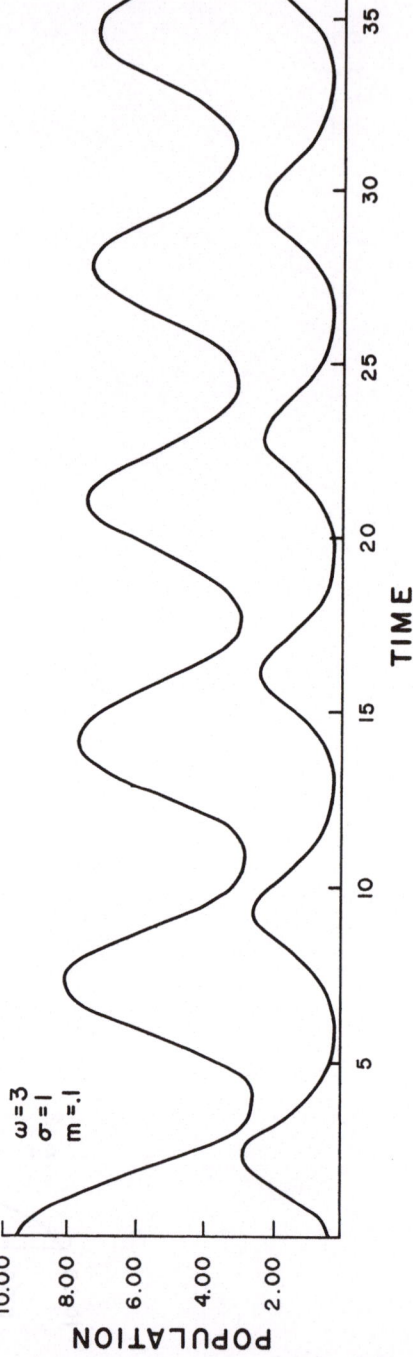

An investigation of equations (5.6) and (5.7) and their interpretation in various models appears in

> K. L. Cooke and J. A. Yorke, Equations Modelling
> Population Growth and Gonorrhea Epidemiology,
> Math. Biosciences 16(1973), 75-101.

Equations (5.8) and (5.9) appear in

> K. L. Cooke, On an Epidemic Equation with Immi-
> gration, to appear.

This paper makes use of results of

> J. J. Levin and D. E. Shea, On the Asymptotic
> Behavior of the Bounded Solutions of some Inte-
> gral Equations I, II, III, J. Math. Anal. Appl.
> 37(1972), 42-82, 288-326, 537-575.

Equation (5.10) is one of the simplest models to allow for an
incubation period, which, for example, is clearly present in measles.
Equation (5.10) is a special case of equation (8.15), p. 140 of

> N. T. J. Bailey, The Mathematical Theory of Epi-
> demics, Griffin Book Co., 1957,

with $u = \gamma = 0$, $(y(t) = N - x(t)$ there).

See also

> H. E. Soper, The Interpretation of Periodicity in
> Disease Prevalence, J. Roy. Statist. Soc. 92(1929),
> 34-61.

> E. B. Wilson and J. Worcester, Damping of Epidemic
> Waves, Proc. Nat. Acad. Sci. 31(1945), 294-298.

> E. B. Wilson and M. Burke, The Epidemic Curve, Proc.
> Nat. Acad. Sci. 38(1942), 361-367.

This equation was also the subject of very early investigations in differential-difference equations; for example,

E. M. Wright, A Nonlinear Difference Differential Equation, J. Reine Angew. Math. 194(1955), 68-74.

The Nonlinear Difference Differential Equation, Quart. J. Math. Oxford Ser. 17(1946), 245-252.

A Functional Equation in the Heuristic theory of primes, Math. Gazette 45(1961), 15-16.

S. Kakutani and L. Markus, On the Nonlinear Difference Differential Equation $y'(t) = (A-B(t-\tau))y(t)$. Contributions to the Theory of Nonlinear Oscillations IV S. (Lefschetz, ed.), Princeton, 1958.

In

F. Hoppensteadt and P. Waltman, A Problem in the Theory of Epidemics II, Math. Biosciences 12(1971), 133-145,

an iterative technique was used to compute solutions of (5.1) - (5.3).

Figures (5.1) - (5.4) were taken from graphs computed by J. Mosevich using the numerical technique outlined in this section. More detailed computations and an explanation of this technique are given in

J. Mosevich, A Numerical Method for Approximating Solutions to the Functional Equations Arising in the Epidemic Models of Hoppensteadt and Waltman, in preparation.

The general subject of numerical solution of functional

differential equations is discussed in the survey article

C. W. Cryer, Numerical Methods for Functional Differential Equations, Delay and Functional Differential Equations and their Application, K. Schmitt,
ed., Academic Press, New York, 1972, 17-101.

6. A TWO POPULATION THRESHOLD MODEL

We consider now a more complicated model involving two populations and an infection which is communicated between them. An
example would be males and females as the populations and a venereal
disease, or mosquito and man as the populations and malaria as the
disease.

Each population is divided into the four classes, which we
denote respectively by $(S)(E)(I)(R)$ and $(\tilde{S})(\tilde{E})(\tilde{I})(\tilde{R})$, as in the
model of Section 4. The infection is presumed to spread according to
the following rules (parentheses indicate the second population)
where $S(t)$ indicates the number of susceptibles of the first population at time t; $\tilde{S}(t)$, the second population; etc:

   (i)   The rate of exposure of susceptibles to infec-
         tives at time t is given by $-r(t)S(t)\tilde{I}(t)$
         $(-\tilde{r}(t)\tilde{S}(t)I(t))$,

   (ii) An individual who is first exposed at time τ
         becomes infective at time t if

$$\int_{\tau}^{t} [\rho_1(x) + \rho_2(x)\tilde{I}(x)]dx = m$$

$$\left(\int_{\tilde{\tau}}^{t} [\tilde{\rho}_1(x) + \tilde{\rho}_2(x)I(x)]dx = \tilde{m} \right)$$

         where $\rho_1(x)$, $\rho_2(x)$, $\tilde{\rho}_1(x)$, $\tilde{\rho}_2(x)$ are nonnega-
         tive functions,

   (iii) An individual who becomes infective at time t
         recovers from the infection at time $t + \sigma$ $(t + \tilde{\sigma})$

where σ, $\tilde{\sigma}$ are given positive constants,

(iv) An individual who recovers from the infection at
time t is immune until time $t+\omega$ $(t+\tilde{\omega})$, at
which time he becomes susceptible again, where
ω, $\tilde{\omega}$ are nonnegative constants,

(v) Each population is constant with

$$I(t) + S(t) + E(t) + R(t) = N \quad (\tilde{I}(t) + \tilde{S}(t) + \tilde{E}(t) + \tilde{R}(t) = \tilde{N}).$$

These are the same hypotheses as the model in Section 4 except
that the infection is spread from infectives in the first population
to susceptibles in the second; from infectives in the second to susceptibles in the first. The equations can be derived exactly as in
Section 4 except that where infectives appear, it will be infectives
of the other population. The model as presented is entirely sym-
metric, only the constants differ. This would be the case, for
example, in the spread of a venereal disease. However, different
assumptions could be made in each population. In parasite-born in-
fections, $\sigma = +\infty$ is a common assumption, for example. One could also
derive a model using more than two distinct populations, or one could
also make a model where the infection spreads due to infectives within
its own population as well as infectives in the other.

As initial conditions, it will be assumed that S_0, $\tilde{S}_0$ are
given constants and $I_0(t)$, $\tilde{I}_0(t)$ are two given monotone increasing
functions, defined on $-\sigma \le t \le 0$, $-\tilde{\sigma} \le t \le 0$, respectively, with
$I_0(-\sigma) = \tilde{I}_0(-\tilde{\sigma}) = 0$. $\tilde{I}_0(t)$, $-\sigma \le t \le 0$ is the "past history" of the
$I_0(0)$ infectives inserted into the susceptible population at time
$t = 0$. (Similarly for $I_0(t)$.) These functions will be extended (as
in Section 4) to the real line by

$$I_0(t) = \begin{cases} 0, & |t| \geq \sigma \\ I(0) - I_0(t-\sigma), & 0 \leq t \leq \sigma, \end{cases}$$

$$\tilde{I}_0(t) = \begin{cases} 0, & |t| \geq \tilde{\sigma} \\ \tilde{I}(0) - \tilde{I}_0(t-\tilde{\sigma}), & 0 \leq t \leq \tilde{\sigma}. \end{cases}$$

If in addition to the above conditions there is a $\tilde{t}_0 < \sigma$ such that

$$\int_0^{\tilde{t}_0} [\tilde{\rho}_1(x) + \tilde{\rho}_2(x)I_0(x)]dx = \tilde{m}.$$

$I_0(t)$ is said to be **admissible**; similarly for $\tilde{I}_0(t)$, if $t_0 < \tilde{\sigma}$, and

$$\int_0^{t_0} [\rho_1(x) + \rho_2(x)\tilde{I}_0(x)]dx = m.$$

Proceeding as before, the model leads to the following ten non-linear equations for τ, $\tilde{\tau}$, I, $\tilde{I}$, S, $\tilde{S}$, E, $\tilde{E}$, R, and $\tilde{R}$.

(6.1)
$$\int_{\tau(t)}^t [\rho_1(x) + \rho_2(x)\tilde{I}(x)]dx = m$$

$$\tau(t) = 0, \quad t \leq t_0$$

(6.2)
$$S(t) = I_1(t) + S_0 - \int_{\tau(t-\sigma-\omega)}^t (r\tilde{I}S)(x)dx$$

$$I_1(t) = \begin{cases} 0, & , \quad t \leq \omega \\ I_0 - I_0(t-\omega), & \omega \leq t \end{cases}$$

$$(6.3) \qquad E(t) = \int_{\tau(t)}^{t} (r\tilde{I}S)(x)dx$$

$$(6.4) \qquad I(t) = I_0(t) + \int_{\tau(t-\sigma)}^{\tau(t)} (r\tilde{I}S)(x)dx$$

$$(6.5) \qquad R(t) = I_2(t) + \int_{\tau(t-\sigma-w)}^{\tau(t-\sigma)} (r\tilde{I}S)(x)dx$$

$$I_2(t) = \begin{cases} I_0 - I_0(t) & , \quad t \le w \\ I_0(t-w) - I_0(t), & w \le t \end{cases}$$

$$(6.6) \qquad \int_{\tilde{\tau}(t)}^{t} [\tilde{\rho}_1(x) + \tilde{\rho}_2(x)I(x)]dx = \tilde{m}$$

$$\tilde{\tau}(t) = 0, \quad t \le \tilde{t}_0$$

$$(6.7) \qquad \tilde{S}(t) = \tilde{I}_1(t) + \tilde{S}_0 - \int_{\tilde{\tau}(t-\tilde{\sigma}-\tilde{w})}^{t} (\tilde{r}\tilde{I}\tilde{S})(x)dx$$

$$\tilde{I}_1(t) = \begin{cases} 0 & , \quad t \le \tilde{w} \\ \tilde{I}_0 - \tilde{I}_0(t-\tilde{w}), & t \ge \tilde{w} \end{cases}$$

$$(6.8) \qquad \tilde{E}(t) = \int_{\tilde{\tau}(t)}^{t} \tilde{r}\tilde{I}\tilde{S}(x)dx$$

$$(6.9) \qquad \tilde{I}(t) = \tilde{I}_0(t) + \int_{\tilde{\tau}(t-\tilde{\sigma})}^{\tilde{\tau}(t)} (\tilde{r}\tilde{I}\tilde{S})(x)dx$$

$$(6.10) \qquad \tilde{R}(t) = \tilde{I}_2(t) + \int_{\tilde{\tau}(t-\tilde{\sigma}-\tilde{\omega})}^{\tilde{\tau}(t-\tilde{\sigma})} (\tilde{\tau}\tilde{I}\tilde{S})(x)dx$$

$$\tilde{I}_2(t) = \begin{cases} \tilde{I}_0 - \tilde{I}_0(t) & , \quad t \le \tilde{\omega} \\ \tilde{I}_0(t-\tilde{\omega}) - \tilde{I}_0(t), & \tilde{\omega} \le t. \end{cases}$$

The initial conditions are:

$$I(t) = I_0(t), \quad -\sigma \le t \le 0$$

$$\tilde{I}(t) = \tilde{I}_0(t), \quad -\tilde{\sigma} \le t \le 0$$

$$S(0) = S_0 = N - I_0(0)$$

$$\tilde{S}(0) = \tilde{S}_0 = \tilde{N} - \tilde{I}_0(0)$$

$$R(0) = \tilde{R}(0) = E(0) = \tilde{E}(0) = 0.$$

Since $(6.3)(6.5)(6.8)(6.10)$ can be solved by an integration if τ, $\tilde{\tau}$, I, $\tilde{I}$, S, $\tilde{S}$ are known, it is really necessary to consider only the remaining six equations.

We note first that if neither initial function is admissible then we can find a solution by defining

$$\tau(t) \equiv 0, \quad \tilde{\tau}(t) \equiv 0$$

$$I(t) = I_0(t), \quad \tilde{I}(t) = \tilde{I}_0(t),$$

and then solving the linear integral equations

$$S(t) = I_1(t) + S_0 - \int_0^t r(x)\tilde{I}_0(x)S(x)dx$$

$$\tilde{S}(t) = \tilde{I}_1(t) + \tilde{S}_0 - \int_0^t \tilde{r}(x)I_0(x)\tilde{S}(x)dx.$$

Of course for $t \geq \max(\sigma,\tilde{\sigma})$ the integrands are identically zero. Thus without loss of generality the admissibility of either $I_0(t)$ or $\tilde{I}_0(t)$ can be assumed. In the case $\rho_1(x) > 0$, $\tilde{\rho}_1(x) > 0$ (the case of an incubation period), the theorem is exactly as in Section 4.

THEOREM 6.1. Let one of $I_0(t)$, $\tilde{I}_0(t)$ be admissible and the other monotone, continuous, and equal to zero at $t = -\sigma$ or $t = -\tilde{\sigma}$ as appropriate, let $r > 0$, $\tilde{r} > 0$, $\rho_1 > 0$, $\tilde{\rho}_1 > 0$, $\rho_2 \geq 0$, $\tilde{\rho}_2 \geq 0$ be continuous functions, and let $\sigma > 0$, $\tilde{\sigma} > 0$, $m \geq 0$, $\tilde{m} \geq 0$, $\omega \geq 0$, $\tilde{\omega} \geq 0$ be given constants. Then there exists a unique continuous solution of (6.1) - (6.10) on $[0,\infty)$ which depends continuously on the initial conditions $I_0(t)$, $\tilde{I}_0(t)$, S_0, $\tilde{S}_0$ and on given functions and parameters in the equations.

On the other hand, if $\rho_1 \geq 0$, $\tilde{\rho}_1 \geq 0$, $\rho_2 > 0$, $\tilde{\rho}_2 > 0$ (threshold case) we have:

THEOREM 6.2. Let $I_0(t)$ and $\tilde{I}_0(t)$ be admissible, $I_0(\tilde{t}_0) > 0$, $\tilde{I}_0(t_0) > 0$, let $r > 0$, $\tilde{r} > 0$, $\rho_1 \geq 0$, $\tilde{\rho}_1 \geq 0$, $\rho_2 > 0$, $\tilde{\rho}_2 > 0$ be continuous functions, and let $\sigma > 0$, $\tilde{\sigma} > 0$, $m \geq 0$, $\tilde{m} \geq 0$, $\omega \geq 0$, $\tilde{\omega} \geq 0$ be given constants. Then there exists a unique continuous solution of (6.1) - (6.10) which depends continuously on the initial conditions $I_0(t)$, S_0, $\tilde{I}_0(t)$, $\tilde{S}_0$ and on the given functions and parameters in the equations. Moreover, $S(t)$, $\tilde{S}(t)$, $I(t)$, $\tilde{I}(t)$ are positive functions.

We prove Theorem 6.1. Note that the proof of a similar theorem in Section 4 was omitted. The proof of Theorem 6.2 is indicated by showing where differences occur from the proof of Theorem 6.1. A better result than Theorem 6.2 is possible at the expense of a more complicated threshold condition. This is discussed after the proof of Theorem 6.2.

Proof of Theorem 6.1. Suppose that t_0 exists and if $\tilde{t}_0$ exists, $t_0 \leq \tilde{t}_0$. Then on $[0,t_0]$ a solution may be found by the procedure described preceding the statement of the theorem. We wish to extend this solution to a larger interval.

Suppose that nonnegative functions $I(t)$, $\tilde{I}(t)$, $S(t)$, $\tilde{S}(t)$, $\tau(t)$, $\tilde{\tau}(t)$ have been found on $[0,t_1]$ with $\tau(t)$ and $\tilde{\tau}(t)$ non-decreasing. We wish to find a solution on $[0,t_2]$, $t_2 > t_1$. Let $C[a,b]$ denote the continuous functions on $[a,b]$ with uniform norm $\|\cdot\|$. Let

$$\mathcal{M} = [\varphi \mid \varphi \in C[t_1,t_2], \ 0 \leq \varphi(t) \leq \tilde{N}, \ \varphi(t_1) = \tilde{I}(t_1)]$$

and define a metric on $\mathcal{M}$ by $d(\varphi_1,\varphi_2) = \|\varphi_1 - \varphi_2\|$. Define $U : \mathcal{M} \to C[t_1,t_2]$ by

$$(6.11) \qquad \int_{U\varphi(t)}^{t_1} [\rho_1(x) + \rho_2(x)\tilde{I}(x)]dx + \int_{t_1}^{t} [\rho_1(x) + \rho_2(x)\varphi(x)]dx = m$$

where $t_2 - t_1$ will be restricted so that $U\varphi(t) \leq t_1$.

LEMMA 6.1. If $t_2 < t_1 + \dfrac{m}{\rho_1^* + \rho_2^*\tilde{N}}$, $U\varphi(t) < t_1$, U satisfies a Lipschitz condition, and $U\varphi(t)$ is monotone nondecreasing in t.

The $*$ notation is as in Section 4, $f^* = \max_I f(t)$, $f_* = \min_I f(t)$, where I is an appropriate closed interval (obvious from the context).

Proof of Lemma 6.1. From the defining equation (6.11)

$$(t - U\varphi(t))(\rho_1^* + \rho_2^* N) \geq m$$

or

$$U\varphi(t) \leq t - \frac{m}{\rho_1^* + \rho_2^* N} < t_1.$$

If $\varphi_1, \varphi_2 \in \mathfrak{m},$ then

$$\int_{U\varphi_1(t)}^{t_1} [\rho_1(x) + \rho_2(x)\tilde{I}(x)]dx + \int_{t_1}^{t} [\rho_1(x) + \rho_2(x)\varphi_1(x)]dx$$

$$= \int_{U\varphi_2(t)}^{t_1} [\rho_1(x) + \rho_2(x)\tilde{I}(x)]dx + \int_{t_1}^{t} [\rho_1(x) + \rho_2(x)\varphi_2(x)]dx$$

or

$$\int_{t_1}^{t} \rho_2(x)(\varphi_1(x) - \varphi_2(x))dx = \int_{U\varphi_2(t)}^{U\varphi_1(t)} [\rho_1(x) + \rho_2(x)\tilde{I}(x)]dx.$$

Hence it follows that

$$\rho_2^* d(\varphi_1, \varphi_2) |t_2 - t_1| \geq \rho_{1*} \|U\varphi_1 - U\varphi_2\|$$

or

(6.12) $$\|U\varphi_1 - U\varphi_2\| \leq \frac{\rho_2^*}{\rho_{1*}} |t_1 - t_2| d(\varphi_1, \varphi_2).$$

Further $U\varphi(t)$ is monotone since $\rho_2(x)\varphi(x) \geq 0$ and $\rho_1(x) > 0$.

On the range of U, denoted $\mathcal{R}U$, define $V : \mathcal{R}U \to C[t_1, t_2]$ by

$$V\varphi(t) = I_0(t) + \int_{\tau(t-\sigma)}^{\varphi(t)} r(x)\tilde{I}(x)S(x)dx.$$

LEMMA 6.2. V satisfies a Lipschitz condition, $N \geq V\varphi(t) \geq 0$ and $V\varphi(t_1) = I(t_1)$.

Proof: First of all,

$$(V\varphi)(t) \leq I_0(t) + \int_{\tau(\varphi(t)-\sigma-w)}^{\varphi(t)} r(x)\tilde{I}(x)S(x)dx$$

$$\leq I_0(t) + I_1(\varphi(t)) + S_0 - S(\varphi(t))$$

$$\leq N.$$

The last inequality follows from the definition of I_1 and the monotonicity of $I_0(t)$. Further, $\varphi(t) \geq \tau(t_1) \geq \tau(t_2-\sigma) \geq \tau(t-\sigma)$ for $t \in [t_1, t_2]$ provided we choose $t_2 - t_1 < \sigma$. Hence $(V\varphi)(t) \geq 0$. If φ_1 and $\varphi_2 \in \mathcal{R}U$, then

$$|V\varphi_1(t) - V\varphi_2(t)| = \left|\int_{\varphi_1(t)}^{\varphi_2(t)} r(x)\tilde{I}(x)S(x)dx\right| \leq r^*N\tilde{N}\|\varphi_2 - \varphi_1\|,$$

or

(6.13) $$\|V\varphi_1 - V\varphi_2\| \leq r^*N\tilde{N}\|\varphi_2 - \varphi_1\|.$$

Since $\varphi(t_1) = \tau(t_1)$, $V\varphi(t_1) = I(t_1)$.

On the range of V, RV, define $\tilde{U} : RV \to C[t_1,t_2]$ as follows: For $t \in [t_1,t_2]$ and $\varphi \in RV$ if there exists a number η such that

$$\int_\eta^{t_1} [\tilde{\rho}_1(x) + \tilde{\rho}_2(x)I(x)]dx + \int_{t_1}^t [\tilde{\rho}_1(x) + \tilde{\rho}_2(x)\varphi(x)]dx = \tilde{m}$$

then $(\tilde{U}\varphi)(t) = \eta$. If no such η exists, set $(\tilde{U}\varphi)(t) = 0$.

LEMMA 6.3. $\tilde{U}\varphi(t) < t_2$, if $t_2 < t_1 + \dfrac{\tilde{m}}{\rho_1^* + \rho_2^* N}$, $\tilde{U}$ satisfies a Lipschitz condition, and $(\tilde{U}\varphi)(t)$ is monotone nondecreasing in t.

Proof: The monotonicity follows since $\tilde{\rho}_1 > 0$, $\tilde{\rho}_2 \geq 0$. Hence if $\tilde{U}\varphi(t_2) \neq 0$, $\tilde{\rho}_1^* + \tilde{\rho}_2^* N(t_2 - \tilde{U}\varphi(t_2)) \geq \tilde{m}$ or

$$\tilde{U}\varphi(t_2) \leq t_2 - \frac{\tilde{m}}{\rho_1^* + \rho_2^* N} < t_1.$$

Suppose $\|\tilde{U}\varphi_1\| = 0$ and $\|U\varphi_2\| \neq 0$. For $t \in [t_1,t_2]$, such that $U\varphi_2(t) \neq 0$ (the other cases follow easily), then

$$\int_0^{t_1} [\tilde{\rho}_1(x) + \tilde{\rho}_2(x)I(x)]dx + \int_{t_1}^t [\tilde{\rho}_1(x) + \tilde{\rho}_2(x)\varphi_1(x)]dx$$

$$\leq \int_{\tilde{U}\varphi_2(t)}^{t_1} [\tilde{\rho}_1(x) + \tilde{\rho}_2(x)I(x)]dx + \int_{t_1}^t [\tilde{\rho}_1(x) + \tilde{\rho}_2(x)\varphi_2(x)]dx$$

or

$$\int_0^{\tilde{U}\varphi_2(t)} \tilde{\rho}_1(x)dx \leq \int_{t_1}^t \tilde{\rho}_2(x) |\varphi_1(x) - \varphi_2(x)|dx$$

and hence it follows that

$$(6.14) \qquad \|\tilde{U}\varphi_2 - \tilde{U}\varphi_1\| \leq \frac{\tilde{\rho}_2^*}{\tilde{\rho}_{1*}}\|\varphi_1 - \varphi_2\| |t_2 - t_1|.$$

Define $\tilde{V} : \mathcal{R}\tilde{U} \to C[t_1, t_2]$ by

$$\tilde{V}\varphi(t) = \tilde{I}_0(t) + \int_{\tilde{\tau}(t-\tilde{\sigma})}^{\varphi(t)} \tilde{r}(x)\tilde{S}(x)I(x)dx.$$

LEMMA 6.4. $\tilde{V}$ <u>satisfies a Lipschitz condition</u> <u>and</u> $\mathcal{R}\tilde{V} \subset \mathcal{M}$.

Proof: If $\varphi_1, \varphi_2 \in \mathcal{R}\tilde{U}$, then

$$|\tilde{V}\varphi_1(t) - \tilde{V}\varphi_2(t)| \leq |\int_{\varphi_1(t)}^{\varphi_2(t)} \tilde{r}(t)\tilde{S}(x)I(x)| \leq \tilde{r}^*N\tilde{N}\|\varphi_2 - \varphi_1\|$$

or

$$(6.15) \qquad \|\tilde{V}\varphi_1 - \tilde{V}\varphi_2\| \leq \tilde{r}^*\tilde{N}N\|\varphi_2 - \varphi_1\|.$$

Further,

$$0 \leq \tilde{V}\varphi(t) \leq \tilde{I}_0(t) + \int_{\tilde{\tau}(\varphi(t)-\tilde{\sigma}-\tilde{w})}^{\varphi(t)} \tilde{r}(x)\tilde{S}(x)I(x)dx$$

$$\leq \tilde{I}_0(t) + \tilde{I}_1(\varphi(t)) + \tilde{S}_0 - \tilde{S}(\varphi(t))$$

$$\leq \tilde{N}.$$

Finally, define $T : \mathcal{M} \to \mathcal{M}$

$$T\varphi = \tilde{V}[\tilde{U}[V[U\varphi]]].$$

Using the Lipschitz conditions (6.12)(6.13)(6.14)(6.15) for each of the four mappings in the composite, one finds

$$\|T\varphi_1 - T\varphi_2\| \le \tilde{r}^* r^* N^2 \tilde{N}^2 \; \frac{\tilde{\rho}_2^* \rho_2^*}{\tilde{\rho}_{1*} \rho_{1*}} \; |t_2 - t_1|^2 \|\varphi_1 - \varphi_2\|.$$

Thus for $|t_2 - t_1|$ sufficiently small T is a contraction. Of course, $|t_2 - t_1|$ must also be small enough to satisfy the other conditions in the lemmas. Since $\mathcal{m}$ is a complete metric space there is a unique fixed point, i.e., $\varphi \in \mathcal{m}$ such that $T\varphi = \varphi$.

Define for $t_1 \le t \le t_2$

$$\tau(t) = U\varphi(t)$$

$$I(t) = V(U\varphi)(t)$$

$$\tilde{\tau}(t) = \tilde{U}VU\varphi(t)$$

$$\tilde{I}(t) = \varphi(t).$$

This provides an extension to $[0, t_2]$ of the solutions of (6.1)(6.4)(6.6)(6.9) given of $[0, t_1]$. $S(t)$ and $\tilde{S}(t)$ can be found as solutions of the linear integral equations.

$$\tilde{S}(t) = \tilde{I}_1(t) + \tilde{S}_0 - \int_{\tilde{\tau}(t-\tilde{\sigma}-\tilde{w})}^{t} rIS(x)dx$$

$$S(t) = I_1(t) + S_0 - \int_{\tau(t-\sigma-w)}^{t} rIS(x)dx.$$

The remaining functions $E(t)$, $\tilde{E}(t)$, $R(t)$, $\tilde{R}(t)$ can be solved when the above six are known.

The above extension procedure can be repeated on a larger

interval $[0,t_3]$, and then to $[0,t_4]$, etc. If these extensions do not cover the positive half of the real line then there is an interval $[0,T)$ beyond which extension is not possible. Since τ and $\tilde{\tau}$ are nondecreasing, $\lim_{t \to T} - \tau(t) = \tau(T^-)$ and $\tilde{\tau}(T^-)$ exist. If these exist, $I(T^-)$, $\tilde{I}(T^-)$ also exist. The above extension procedure allows $\tau(t)$, $\tilde{\tau}(t)$, $I(t)$, $\tilde{I}(t)$ to be extended to $[0,T+\epsilon]$. Hence $S(t)$, $\tilde{S}(t)$ can also be extended, which contradicts the maximality of T. Thus we have a solution on $[0,\infty)$.

Since the solution on $[0,t_0]$ was unique, and since also each extension was unique, the solutions are unique. Proof of the continuous dependence statements is omitted.

Proof of Theorem 6.2: The proof proceeds along the same lines as Theorem 6.1, using the basic set as in Theorem 4.1. We sketch the proof, mainly indicating where differences occur in the sequence of lemmas. The initial solution is as before since both $I_0(t)$ and $\tilde{I}_0(t)$ are admissible and we consider only the extension procedure. Let

$$\mathcal{M} = \{\varphi \mid \varphi \in C[t_1,t_2], \ \gamma \le \varphi(t) \le \tilde{N}, \ \varphi(t_1) = \tilde{I}(t_1)\}$$

where $\gamma = \tilde{I}_0(t_0)/2$ if $t_1 = t_0$ and $\gamma = \tilde{r}_* I_* \tilde{S}_* [\tilde{\tau}(t_1) - \tilde{\tau}(t_2 - \tilde{\sigma})]$ if $t_1 > t_0$, and we assume a priori $t_2 - t_1 < \min(\sigma, \tilde{\sigma})$. If $t_1 > t_0$, τ is strictly increasing so $\gamma > 0$. Lemma 6.1 is as before except that the estimate (6.12) becomes

$$(6.16) \qquad \frac{\rho_2^* d(\varphi_1, \varphi_2)}{\rho_{1*} + \rho_{2*}\gamma} \ |t_1 - t_2| \ge \|U\varphi_1 - U\varphi_2\|.$$

The estimate in Lemma 6.2 is exactly the same, i.e.,

(6.17)
$$\|V\varphi_1 - V\varphi_2\| \le r^* N \tilde{N} \|\varphi_1 - \varphi_2\|.$$

However, a positive lower bound on the range is now important. If $t_1 \le \tilde{t}_0$ then restrict t_2 to be so small that $\alpha = I_0(\tilde{t}_0)/2 > 0$. If $t_1 > \tilde{t}_0$,

$$V\varphi(t) \ge \int_{\tau(t-\sigma)}^{\varphi(t)} r(x)\tilde{I}(x)S(x)dx$$

$$\ge r_* \tilde{I}_* S_* (\varphi(t) - \tau(t-\sigma))$$

$$\ge r_* \tilde{I}_* S_* (\tau(t_1) - \tau(t_2-\sigma)) = \alpha > 0.$$

The inequality $\alpha > 0$ follows since τ is strictly monotone and $r_* \tilde{I}_* S_* > 0$. The upper bound follows as before so

$$\alpha \le V\varphi(t) \le N.$$

That $\alpha > 0$ makes $\tilde{U}\varphi$ in Lemma 6.3 strictly increasing in t. The upper bound is as before. Since now $\rho_2 > 0$ (and perhaps $\rho_1 \equiv 0$), the Lipschitz estimate changes slightly. From

$$\left| \int_{\tilde{U}\varphi_1(t)}^{\tilde{U}\varphi_2(t)} [\tilde{\rho}_1(x) + \tilde{\rho}_2(x)I(x)]dx \right| \le \int_{t_1}^{t} \tilde{\rho}_2(x) |\varphi_2(x) - \varphi_1(x)| dx,$$

it follows that

$$(\tilde{\rho}_{1*} + \tilde{\rho}_{2*}(x)I_*) |\tilde{U}\varphi_2(t) - \tilde{U}\varphi_1(t)| \le \tilde{\rho}_2^* \|\varphi_2 - \varphi_1\| |t_2 - t_1|,$$

or

$$(6.18) \qquad \|\tilde{U}\varphi_2 - \tilde{U}\varphi_1\| \le \frac{\tilde{p}_2^*}{\tilde{p}_{1*} + \tilde{p}_{2*}I_*} |t_2 - t_1| \|\varphi_2 - \varphi_1\|.$$

In Lemma 6.4 the Lipschitz argument and the upper bound carry over. To map back into $\mathcal{M}$ it is necessary to establish a lower bound on the range of $\tilde{V}$.

$$(\tilde{V}\varphi)(t) \ge \tilde{I}_0(t) + \tilde{r}_* \tilde{S}_* I_* [\tilde{\tau}(t_1) - \tilde{\tau}(t_2 - \sigma)]$$

If $t_1 = t_0$, t_2 was chosen so that $\tilde{I}_0(t) > I_0(t_0)/2$ for $t_1 \le t \le t_2$. If $t > t_1$, $\tilde{\tau}(t_1) - \tilde{\tau}(t_2 - \tilde{\sigma})) > 0$ and in either case

$$(\tilde{V}\varphi)(t) \ge \gamma,$$

and hence $\mathcal{R}\tilde{V} \subset \mathcal{M}$.

Since $\mathcal{M}$ is a complete metric space, there is a unique fixed point and hence a unique solution on $[0, t_2]$. The extension argument is exactly as before except that if $[0,T)$ is the alleged maximal interval one must show $S(T^-) > 0$, $\tilde{S}(T^-) > 0$. These facts follow, as in Section 4, from the inequalities

$$S'(t) \ge -r(t)S(t)\tilde{I}(t), \quad \text{a.e.}$$

$$\tilde{S}'(t) \ge -\tilde{r}(t)\tilde{S}(t)I(t), \quad \text{a.e.}$$

This completes the proof.

In order not to assume both $I_0(t)$ and $\tilde{I}_0(t)$ are admissible it is necessary, as noted before, to express a condition on the relation between one of the initial infective populations and the threshold. We do this (and summarize the other conditions) in the following condition. If both $I_0(t)$ and $\tilde{I}_0(t)$ are admissible it is obviously satisfied.

The pair $I_0(t)$, $\tilde{I}_0(t)$ is said to satisfy condition (A) if

(i) $I_0(t)$, $\tilde{I}_0(t)$ are monotone nondecreasing, on $[-\sigma,0]$ and $[-\tilde{\sigma},0]$, respectively,

(ii) $I_0(-\sigma) = \tilde{I}_0(-\tilde{\sigma}) = 0$,

(iii) There exists $t_0 < \tilde{\sigma}$ such that

$$\int_0^{t_0} [\rho_1(x) + \rho_2(x)\tilde{I}_0(x)]dx = m,$$

(iv) There exists η, $t_0 < \eta < t_0 + \sigma$ such that

$$\int_0^{\eta} [\rho_1(x) + \rho_2(x)I_0(x)]dx$$

$$+ S_0\int_{t_0}^{\eta} \rho_2(x)\left[1 - \exp - \int_0^{R^{-1}(R(x)-m)} r(u)\tilde{I}_0(u)du\right]dx > \tilde{m}$$

where

$$R(t) = \int_0^t [\rho_1(x) + \rho_2(x)\tilde{I}_0(x)]dx.$$

We illustrate condition (iv) by an example. Let $\omega = \tilde{\omega} = r = \tilde{r} = \rho_2 = \tilde{\rho}_2 = 1$, $\rho_1 = \tilde{\rho}_1 = 0$, $S_0 = 10$, $m = \tilde{m} = 1/2$, $\tilde{I}_0 = 1 + \sin t$, $= \pi/2 \le t \le 0$, and $I_0(t) = 0$, $-\sigma \le t \le 0$. For $0 \le t \le \pi/2$, $\tilde{I}_0(t) = 1 - 1 - \sin(t-\pi/2) = \cos t$. Hence $R(t) = \sin t$, $t \ge 0$, $R^{-1}(t) = \arcsin t$, $0 \le t \le 1$. To check the integral condition we note that

$$S_0 \int_{t_0}^{\pi/4} \left[1 - \exp - \int_0^{R^{-1}(R(x)-1/2)} \cos u \ du\right] dx$$

$$= 10 \int_{\pi/6}^{\pi/4} [1 - \exp(1/2 - \sin x)] dx$$

$$\geq 10 \int_{\pi/6}^{\pi/4} [\sin x - 1/2 - (\sin x - 1/2)^2] dx$$

$$= 10 \left[\frac{\pi}{48} - \sqrt{2} + \sqrt{3} + 1/4 - \frac{\sqrt{3}}{8}\right] > \frac{10\pi}{48} > 1/2$$

where we have used $1 - e^{-x} \geq x - x^2$, $x > 0$.

The following is a generalization of Theorem 6.2.

THEOREM 6.3. Suppose $\rho_1(x) \geq 0$, $\tilde{\rho}_1(x) \geq 0$, $\rho_2(x) > 0$, $\tilde{\rho}_2(x) > 0$, $r(x) > 0$, $\tilde{r}(x) > 0$ are continuous functions, $I_0(t)$, $\tilde{I}_0(t)$ are given initial functions which satisfy condition (A) with $\tilde{I}_0(\eta) > 0$, and $m \geq 0$, $\tilde{m} \geq 0$, $\omega \geq 0$, $\tilde{\omega} \geq 0$, $\sigma > 0$, $\tilde{\sigma} > 0$ are given constants. Then there exists a unique continuous solution of equations (1) - (6). The solutions depend continuously on the initial conditions, given functions, and parameters. Further, $\tilde{I}(t)$, $S(t)$, $\tilde{S}(t)$ are positive for all finite t and $I(t)$ is eventually positive.

Proof: On the interval $[0, t_0]$ a solution can be found as before by setting

$$\tau = \tilde{\tau} = 0$$

$$I(t) = I_0(t)$$

$$\tilde{I}(t) = \tilde{I}_0(t)$$

and by finding $S(t)$, $\widetilde{S}(t)$ by solving

$$S(t) = I_1(t) + S_0 - \int_0^t r(x)\widetilde{I}_0(x)S(x)dx$$

$$\widetilde{S}(t) = \widetilde{I}_1(t) + \widetilde{S}_0 - \int_0^t \widetilde{r}(x)I_0(x)\widetilde{S}(x)dx.$$

As long as $\int_0^t \widetilde{\rho}_2(x)I(x)dx < m$, we can take $\widetilde{r}(t) = 0$ and $\widetilde{I}(t), S(t)$, as defined above. We want to obtain estimates on $I(t)$ in terms of the given functions $I_0(t)$, $\widetilde{I}_0(t)$ which show that there is a point $\widetilde{t}_0$ such that

$$\int_0^{\widetilde{t}_0} [\widetilde{\rho}_1(x) + \widetilde{\rho}_2(x)I(x)]dx = \widetilde{m}.$$

As long as such a $\widetilde{t}_0$ does not exist, $\tau(t)$ is defined as the solution of the initial value problem

$$\tau'(t) = \frac{\rho_2(t)\widetilde{I}_0(t) + \rho_1(t)}{\rho_2(\tau)\widetilde{I}_0(\tau) + \rho_1(\tau)}, \qquad t > t_0$$

$$\tau(t_0) = 0,$$

and by $\tau(t) \equiv 0$, $t < t_0$. On this same interval,

$$I(t) = I_0(t) + \int_{\tau(t-\sigma)}^{\tau(t)} r(x)S(x)\widetilde{I}_0(x)dx.$$

Suppose $\widetilde{t}_0$ does not exist. Then one has an immediate estimate on $S(t)$,

(6.19) $$S(t) \geq S_0 \exp - \int_0^t r(x)\widetilde{I}_0(x)dx, \qquad \text{a.e.,}$$

since $I_1'(t) \geq 0$, a.e. Hence on $[0,\tilde{\sigma}]$,

$$I(t) \geq I_0(t) + S_0 \int_0^{\tau(t)} r(x)\tilde{I}_0(x) \exp\left(-\int_0^x r(u)\tilde{I}_0(u)du\right)dx$$

$$= I_0(t) + S_0\left[1 - \exp - \int_0^{\tau(t)} r(u)\tilde{I}_0(u)du\right]$$

or

$$(6.20) \qquad I(t) \geq I_0(t) + S_0\left[1 - \exp - \int_0^{R^{-1}(R(t)-m)} r(u)\tilde{I}_0(u)du\right]$$

Hence, using (6.20) it follows that

$$\int_0^\eta [\rho_1(x) + \rho_2(x)I(x)dx]$$

$$\geq \int_{t_0}^\eta [\rho_1(x) + \rho_2(x)I_0(x)]dx + S_0\int_0^\eta [1 - \exp - \int_0^{R^{-1}(R(x)-m)} r(u)\tilde{I}_0(u)du\,dx$$

$$\geq \tilde{m}.$$

Thus on an interval $[0,t_1]$, $t_0 \leq t_1 \leq \eta < \tilde{\sigma}$, we have a solution $\mathcal{T}(t) = 0$, $\tau(t)$, $I(t)$, $\tilde{I}(t) = I_0(t)$, $S(t)$, $\tilde{S}(t)$, with $I(t_1) > 0$, $\tilde{I}(t_1) > 0$, $S(t_1) > 0$, $\tilde{S}(t_1) > 0$. Since these functions are positive, the extension procedure described in Theorem 6.2 now applies as does the remainder of the proof there.

Notes:

The material in this section appears in

P. Waltman, A deterministic model of the spread

of an infection between two populations,
Delay and Functional Differential Equations
and their Applications, K. Schmitt, Ed.,
Academic Press, New York, 1972, 281-291.

, A threshold criterion for the spread
of an infection in a two population model, Math.
Biosciences, to appear.

A very detailed two population model (not of the form treated
here) and a very specific account of the use of mathematical models
in epidemiology can be found in

G. MacDonald, The Epidemiology and Control of
Malaria, Oxford University Press, London, 1957.

7. A MODEL WITH AGE DEPENDENCE AND AN OPEN POPULATION

The models in the previous sections assumed a constant, closed total population or, from another point of view, birth, death, and migration were in exact balance. In this section a special case of a model of F. Hoppensteadt is described in which these factors are allowed to vary and a further complication is permitted in the model in that age dependent rates of exposure, virility, etc., are also allowed. In fact, two types of age dependence will be introduced, dependence on chronological age and on class age. The first of these allows the disease to selectively infect a particular age group while the second allows the important rate constants to vary with the length of time individuals have been in a given class; for example, allowing the probability of transmitting the disease to decrease when the individual has been infective for a long period of time. The pair (a,c) (chronological age, class age) will be referred to as the age of the individual.

Let (S)(I)(R) be the same classes as in Section 1, susceptible, infective, and removed. Instead of working with the number of individuals in the class we now use densities. For example at time t, $S(a,c,t)$ will be the density of susceptibles of age (a,c). Thus the number of individuals in the susceptible class at time t between the chronological ages of a_1 and a_2 is given by

$$N \int_{a_1}^{a_2} \int_0^\infty S(a,c,t) dc\, da$$

where N is some appropriate scale factor, a is chronological age, c is class age. Hereafter we take $N = 1$.

Concerning the spread of the infection we make the following assumption:

(i) New susceptibles occur in the population through births (by parents in all classes) and through immigration at a given rate. Only infectives die. The birth rates for each class β_i, $i = 1,2,3$, the immigration rate m, and death rate d, are assumed given,

(ii) The infective period lasts (barring death) for a fixed time σ,

(iii) The change in the density of susceptibles of age (a,c) at time t through interaction with infectives is given by

$$S(a,c,t)\int_0^\infty \int_0^\sigma r(a,c,t,a',c')I(a',c',t)dc'\,da'$$

where r is a given function. r expresses the effect of infectives of age (a',c') on susceptibles of age (a,c).

A change of time $h > 0$ results in the advancement of the age (a,c) to (a+h,c+h). Hence the change in the density of susceptibles between t and t +h is given by

$$S(a+h,c+h,t+h) - S(a,c,t).$$

We use this to derive the differential equation which expresses the dynamics of the density S(a,c,t). Immigration into the class S occurs at class age zero and births at age (0,0). The rate of change in S is the exposure rate due to contacts with infectives. This quantity can be expressed as

$$-S(a,c,t)\int_0^\infty \int_0^\sigma r(a,c,t,a',c')I(a',c',t)dc'da'$$

by (iii). Hence the change in S can be described by

(7.1) $$\frac{\partial S}{\partial a} + \frac{\partial S}{\partial c} + \frac{\partial S}{\partial t}$$

$$= -S(a,c,t)\int_0^\infty \int_0^\sigma r(a,c,t,a',c')I(a',c',t)dc'da',$$

$$S(a,0,t) = m(a,t), \quad a > 0,$$

with initial (time) condition

(7.2) $$S(a,c,0) = S_0(a,c)$$

where $S_0(a,c)$ is given. The first equation (7.1) gives the dynamics of $S(a,c,t)$ while the second gives the input of class age zero and positive chronological age. The condition (7.2) expresses the initial age distribution of the susceptible population.

The description of the changes in class (I) is somewhat more delicate. Individuals move into (I) at class age zero and out of (I) to the recovered class (R) only at infective class age, σ. The only other change in the class (I), and the only one possible if the class age, c, satisfies $0 < c < \sigma$, is exit via death. Although in Section 1 "recovered and immune" and "dead" were lumped together in a removed class, if we are to keep track of age it is necessary to distinguish between them. Thus the change in class (I) is described by the equations

$$(7.3) \qquad \frac{\partial I}{\partial a} + \frac{\partial I}{\partial c} + \frac{\partial I}{\partial t} = -d(a,c,t)I$$

$$(7.4) \qquad I(a,0,t) = \int_0^\infty \int_0^\infty \int_0^\sigma r(a,c,t,a',c',)I(a',c',t)S(a,c,t)dc'da'dc$$

with the expected initial condition

$$(7.5) \qquad I(a,c,0) = I_0(a,c).$$

Equation (7.3) describes the dynamics while equation (7.4) describes the redistribution of susceptibles into the infective class. The class (R) is then easy to describe,

$$(7.6) \qquad \frac{\partial R}{\partial a} + \frac{\partial R}{\partial c} + \frac{\partial R}{\partial t} = 0$$

$$(7.7) \qquad R(a,0,t) = I(a,\sigma,t)$$

$$(7.8) \qquad R(a,c,0) = 0.$$

The class age cannot exceed the chronological age so there are also auxiliary conditions,

$$(7.9) \qquad S(a,c,t) = I(a,c,t) = R(a,c,t) = 0, \quad c > a.$$

Since all newly born are in the susceptible class there are also the additional boundary conditions

$$(7.10) \qquad I(0,0,t) = Q(0,0,t) = R(0,0,t) = 0.$$

$$(7.11) \quad S(0,0,t) = \int_0^\infty \int_0^\infty [\beta_1(a,c,t)S(a,c,t) + \beta_2(a,c,t)I(a,c,t)$$

$$+ \beta_3(a,c,t)R(a,c,t)]dc \, da \, .$$

The last equation, (7.11), represents the increase in S as the result of births by parents in S, I, and R (β_1,β_2,β_3 being the respective birth rates). The last two equations follow from the assumption that all newborns are susceptible.

Equations (7.1) - (7.11) constitute the model.

THEOREM: If $r,d,\beta_1, i = 1,2,3, S_0, I_0, R_0$ are nonnegative and continuous then there exist unique nonnegative functions S,I,R satisfying the above equations for $a,c,t > 0$.

This theorem is a special case of a more general theorem for systems of partial differential equations and the reader is referred to the original paper for details (see Notes). The model and the attendant mathematics is obviously much more complex than the ordinary differential equations model of Section 1.

Notes:

The model presented in this section is a special case of one given in

F. Hoppensteadt, An Age Dependent Model, J. Franklin Institute, to appear.

An outline of an existence theorem for a system of partial differential equations which includes this model appears in the appendix of the above paper. Also included is a reduction of the model to an age independent version. This reduction includes the Kermack-McKendrick model of Section 1 as a special case. The introduction of delays and thresholds into the model will be included in a monograph currently being prepared by Professor Hoppensteadt (tentative title, "Some Mathematical Theories of Populations, Genetics, and Epidemics").

8. SOME SIMPLE CONTROL ASPECTS

In this section we consider the problem of attempting to control a predicted epidemic. Suppose, using the current initial state in one of the previously described models, a large outbreak of the infection is predicted. There are two immediate possible courses of action — vaccination and quarantine. In the context of the models, vaccination is taking an individual from susceptible class and putting him in the removed class (or a special class) without passing through the infective stage while quarantine (isolation) is removing an individual from the infective class before the disease has run its course. Only control by vaccination will be treated in this section, but the techniques of solution could be applied in other cases.

The model will be that of Section 1, the simplest model, with the addition of Class (V), the vaccinated individuals. Let $\alpha(t)$ denote the vaccination rate as a function of time. The model then is

$$S'(t) = -rIS - \alpha$$

$$I'(t) = rIS - \gamma I$$

(8.1)
$$R'(t) = \gamma I$$

$$V'(t) = \alpha$$

$$S(0) = S_0, \quad I(0) = I_0, \quad R(0) = V(0) = 0.$$

It is explicitly assumed that only susceptibles are vaccinated and that the inoculation is effective immediately. The first of these presupposes some sort of test for susceptibility before inoculation to prevent "wasted" vaccination of infectives. This could be

corrected by using $\alpha S(t)/N$ (where N is the total population) as
the effect of vaccination on the susceptible class. The methods here
still apply, although some "short cuts" used in the computer program
to account for the class (V) would need to be eliminated. However,
we note that all of the models we have used ignore those individuals
who are naturally immune. On these vaccination is also "wasted" if
given generally. The presumed "test" for susceptibility covers
these while the above modification of the vaccination rate does not.
We are also implicitly assuming that the vaccination rate α is not
applied over a time period sufficient to make $S(t) < 0$. This restriction was "built in" all of the computations. If the vaccine is
not 100% effective, $\alpha(t)$ denotes the effective rate.

If, with the initial conditions (S_0, I_0), the model with
$\alpha(t) \equiv 0$ predicts the spread of the infection through the population
with results that are unacceptable, the course of spread of the
infection is to be altered by introducing a vaccination program,
that is, introducing a nontrivial function $\alpha(t)$ in (8.1). If the
infection causes serious harm to the individuals, it is desirable to
reduce the total number infected to an "acceptable" amount. (The
level of acceptability is a trade-off between expenditure of resources
and discomfort to individuals or damage to society). On the other
hand, it may in some circumstances by important only to limit the
number of individuals infected at any one time. As an example of the
latter case one thinks of a group maintaining an important installation and a minor infection acting in that group. Here the important
consideration may be to keep enough of the population healthy at any
one time so that the installation can be maintained rather than to
limit the total number infected. "Preventing an epidemic" is defined

to be controlling these two quantities: the peak of the number infected at any particular time, and the total number infected in the
given time interval. Of course, doing something to prevent the
spread involves a cost (dollar costs and perhaps social costs). The
control objective is to minimize this cost.

For computational reasons, it is convenient to normalize the
dependent variables and treat the proportion of susceptible, infective, removed, and vaccinated individuals. If we let $\bar{S} = S/N$,
$\bar{I} = I/N$, $\bar{R} = R/N$, $\bar{V} = V/N$, $\bar{\beta} = \beta/N$, $\bar{\alpha} = \alpha/N$, $\bar{S}_0 = S_0/N$, and $\bar{I}_0 = I_0/N$,
then (8.1) becomes

$$\bar{S}' = \overline{rSI} - \bar{\alpha}$$

$$\bar{I}' = \overline{rSI} - \gamma\bar{I}$$

$$\bar{R}' = \gamma\bar{I}$$

$$\bar{V}' = \bar{\alpha}$$

$$\bar{S}(0) = \bar{S}_0, \bar{I}(0) = \bar{I}_0, \bar{R}(0) = \bar{V}(0) = 0$$

where $\bar{S}_0 + \bar{I}_0 = 1$ and $\bar{S}(t) + \bar{I}(t) + \bar{R}(t) + \bar{V}(t) = 1$. The time dimension
could also be scaled but is not for computational reasons
as will be noted below. Since the above system is formally the same
as (8.1) with N normalized to 1, we will use (8.1) with $N = 1$
and think of S, I, R, and V as the proportion of the population
in the various classes. Note that the function $\alpha(t)$ and the parameter r have changed their original meaning and are now dependent on
the total population size, N.

The effort expended in "preventing an epidemic" by vaccination
is assumed to be proportional to the vaccination rate α, i.e.,

achieving a higher vaccination rate means that more equipment, personnel, supplies, etc., must be brought to the locale. Let the expenditure (money, equipment, personnel, supplies, disruption of health care elsewhere, etc.) be described by a (nonlinear) cost functional $C(\alpha)$. $\alpha(t)$ will be restricted to be a step function; for example, α can be thought of as being constant for one time period (a day or a work shift).

The two measures of the size of the epidemic are the total proportion, $R(T) + I(T)$, of the population affected by the infection in the time interval $[0,T]$ and the maximum number infected at the peak, $\max_{t \in [0,T]} I(t)$. The problem can now be stated formally.

PROBLEM: Given a cost functional $C(\alpha)$, positive constants I_0, S_0, A, B, T, and a class of allowable range points of the step functions α, choose a function $\alpha(t)$ such that the solution of (8.1) with initial conditions I_0, S_0, and this $\alpha(t)$, yields a solution such that

(i) $R(T) + I(T) \leq A$,

(ii) $\max_{[0,T]} I(t) \leq B$,

and

$$C(\alpha) = \text{minimum.}$$

In more concrete terms this says that, given the achievable vaccination rates, a cost functional $C(\alpha)$, and the current situation (I_0, S_0), select a pattern of vaccination that will achieve: (i) no more than A of the population succumbs to the infection by time T, and (ii) no more than B of the population is infective at any one time. Moreover, do this so as to minimize the cost.

A solution to this problem can be constructed using the technique of dynamic programming. The procedure was applied to three different epidemics corresponding to $\rho = .3$ (a severe epidemic with unvaccinated intensity $R(\infty) = .96$), $\rho = .6$ (less severe with $R(\infty) = .67$), and $\rho = .8$ (a relatively mild case with $R(\infty) = .37$). r was fixed at $r = 10$ and γ was varied so as to make the epidemic unfold relatively quickly. (This is the reason time was not scaled in the original model.) Ten equally spaced range points (vaccination rates) were allowed. The goals corresponding to the definition of "preventing an epidemic" were taken to be (i) $R(T) + I(T) \leq A$, and (ii) $\max_{n \leq N^*} I(n\Delta) \leq B$ with the values of A and B chosen as indicated in Table 8.1. T was chosen as 1.5 and the interval $[0,1.5]$ was broken into increments Δ of $.1$. Thus there were fifteen vaccination intervals with ten choices of vaccination rate at each interval so that there were 10^{15} possible vaccination profiles (schedules) or step functions $\alpha(t)$.

Let

$$(8.2) \qquad \alpha(t) = \{M_j d \mid j\Delta \leq t \leq (j+1)\Delta, \quad j = 0,1,\cdots,14\}$$

where M_j is an integer between 0 and 9 and the values of d chosen appropriate to the problem (see Table 8.1). The cost function was chosen to be

$$C(\alpha) = \sum_{j=0}^{14} (M_j d\Delta)^2.$$

This choice of cost function reflects a relatively high penalty to achieve high vaccination rates.

TABLE 8.1

Parameters			Goals		Vaccination Profile	Cost	Discretized Problem					Runge-Kutta with No Vaccination				Runge-Kutta with Vaccination				
ρ	lo	d	A	B	Non-zero Entries Only		I(T)	S(T)	R(T)	V(T)	Im	I(T)	S(T)	R(T)	Im	I(T)	S(T)	R(T)	V(T)	Im
.3	.02	.3	.4	.2	3,2,2,2,2,1,2,1	3.15	.03	.10	.36	.51	.16	.0787	.0545	.8668	.3447	.0279	.0645	.3976	.51	.1541
	.04				4,3,3,2,2,1,1,1	4.05	.02	.10	.37	.51	.16	.0606	.0494	.8900	.3510	.0213	.0629	.4058	.51	.1596
	.06				4,4,4,2,1,1,1	4.95	.02	.10	.37	.51	.16	.0516	.0465	.9019	.3571	.0171	.0562	.4167	.51	.1741
	.10				5,5,4,3,2	7.11	.02	.04	.37	.57	.20	.0416	.0425	.9159	.3704	.0086	.0184	.4030	.57	.2072
	.14				9,8,1	13.14	.02	.08	.36	.54	.20	.0358	.0394	.9248	.3838	.0108	.0535	.3956	.54	.2116
.6	.01	.4	.2	.05	1,1,1,1,1,1,1,1,1,1,1,	1.60	.01	.40	.19	.40	.0375	.0485	.3852	.5662	.0995	.0089	.3704	.2201	.40	.0376
	.02				2,2,1,1,1,1,1	2.24	.0075	.40	.1925	.40	.0425	.0315	.3490	.6195	.1056	.0060	.3554	.2348	.40	.0433
	.03				3,3,2,1,1	3.84	.005	.40	.195	.40	.0450	.0235	.3305	.6460	.1118	.0040	.3676	.2278	.40	.0473
	.04				5,4,1	6.72	.005	.40	.195	.40	.0500	.0187	.3175	.6638	.1180	.0041	.3684	.2275	.40	.0528
.8	.01	.2	.1	.05	3,2,2,2,1,2,1	1.08	.0075	.65	.0825	.26	.0125	.0226	.6993	.2781	.0295	.0029	.6304	.1067	.26	.0134
	.015				5,3,2,1	1.72	.0075	.65	.0825	.26	.0175	.0191	.6571	.3238	.0336	.0026	.6160	.1214	.26	.0176
	.02				5,5,4,3,2,1	3.20	.0025	.50	.0975	.40	.0225	.0161	.6238	.3556	.0376	.0006	.4866	.1128	.40	.0229
	.025				7,6,3,1,2,1	4.00	.0025	.50	.0025	.40	.0275	.0137	.6065	.3798	.0417	.0006	.4759	.1236	.40	.0274

The results of the computations are presented in Table 8.1. The optimum "vaccination profile" is represented by the values of M_j in (8.2) starting from time zero. Only the nonzero beginning values of M_j are given, so the vaccination profiles would be completed by adding enough zeros on the right to make 15 values. Under the discretized problem heading, the values given are for the proportion of the population in each class, as found by the dynamic programming solution, at final time T. Since a considerable amount of rounding occurred in construction of the tables required by dynamic programming, a Runge-Kutta program was used to solve the differential equations (8.1) directly using the selected vaccination profile. This provides a check on the accuracy of the alleged solution. The table of values given without vaccination corresponds to the predicted course of the epidemic, and the values with vaccination were determined using the vaccination profile found by dynamic programming. The differences between the discrete problem values and Runge-Kutta with vaccination values give an indication of the inherent error caused by the discretizing or rounding required to make the solution tables for the dynamic programming problem. Making the mesh sizes of the tables finer could reduce this error, but more computations and storage would be required.

To show how to use the data of Table 8.1, we follow one computation completely. For $\rho = .3$ and $I_0 = .06$ the following choice of vaccination rates resulted.

time	rate	% vaccinated
.0 to .3	1.2	36
.3 to .4	.6	6
.4 to .7	.3	9
.7 to 1.5	0	0

To illustrate these results, graphs of the complete Runge-Kutta solution in the vaccinated and the unvaccinated cases for this problem are presented in Figures 8.1 - 8.5. Without vaccination, 95% of the population would have been affected by the disease at time 1.5 (Figure 8.3).

The area discussed in this section seems to be one where further investigations would be profitable. For example, more sophisticated dynamical models (particularly with thresholds) could be used. The vaccination could be applied to the entire population, for it may not be possible to determine who is already infected. Moreover, there may be a delay between inoculation and the point at which immunity is acquired. Dynamic programming fits the model of this section well enough but it was forcibly discretized. Perhaps more sophisticated use of the machinery of control theory would allow more general, and more realistic, problems to be treated.

Notes:

The problem discussed in this section is given in

> H. W. Hethcote and P. Waltman, Optimal Vaccination
> Schedules in a Deterministic Epidemic Model, Math.
> Biosciences, 18(1973), 365-382.

The tables and graphs all appear there. An Appendix gives the details of the application of dynamic programming to this particular problem. Figure 1.1 also appears there.

The ideas used are no doubt common to many problems involving dynamic programming. In particular, they appeared in

FIGURE 8.1

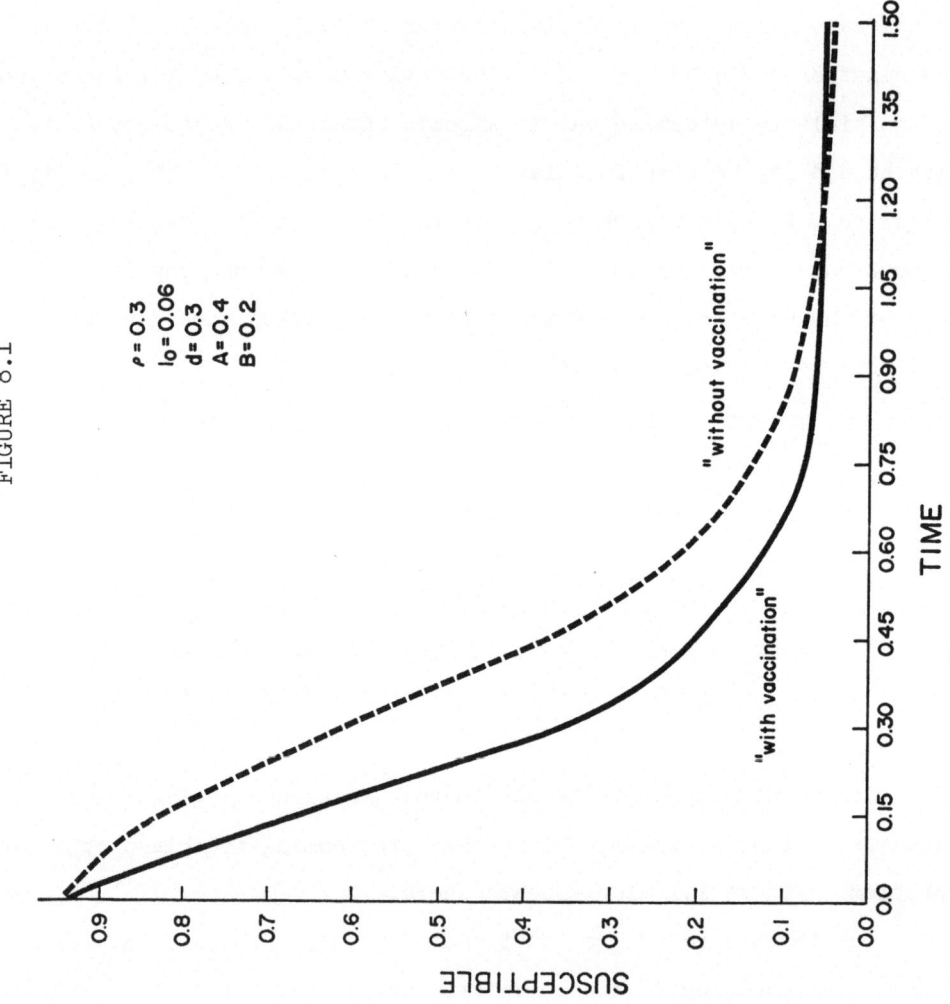

$\rho = 0.3$
$I_0 = 0.06$
$d = 0.3$
$A = 0.4$
$B = 0.2$

"without vaccination"

"with vaccination"

TIME

SUSCEPTIBLE

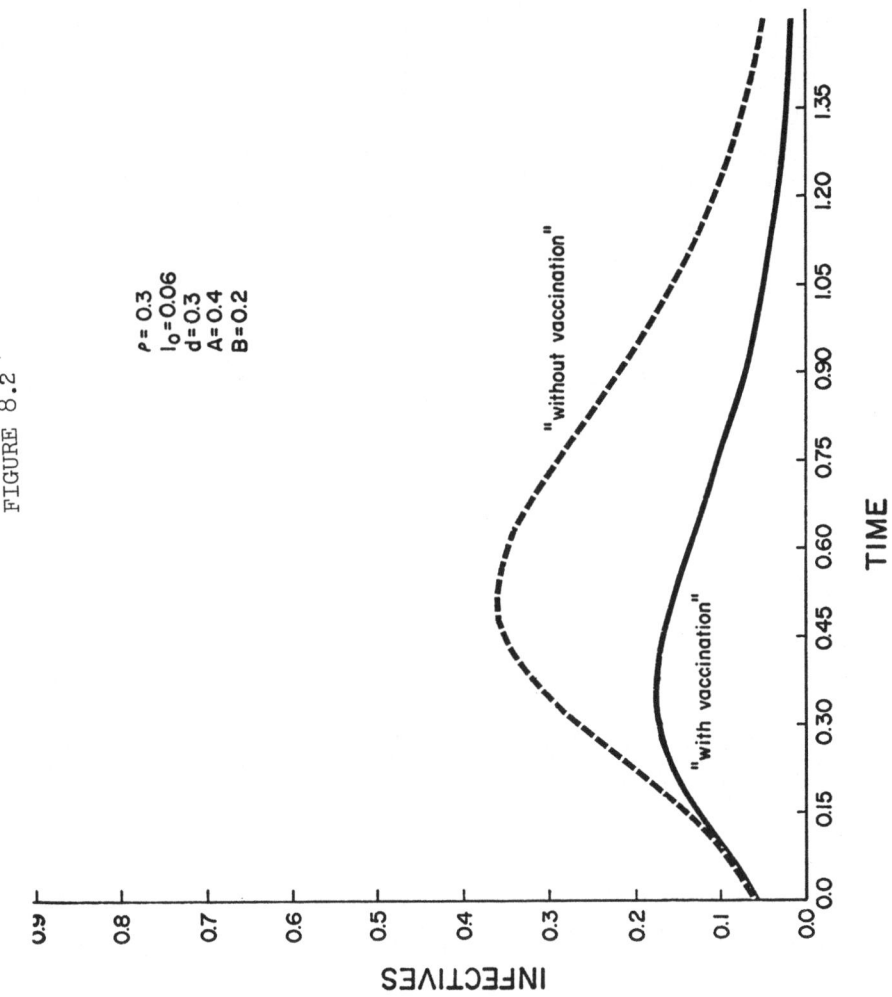

FIGURE 8.2

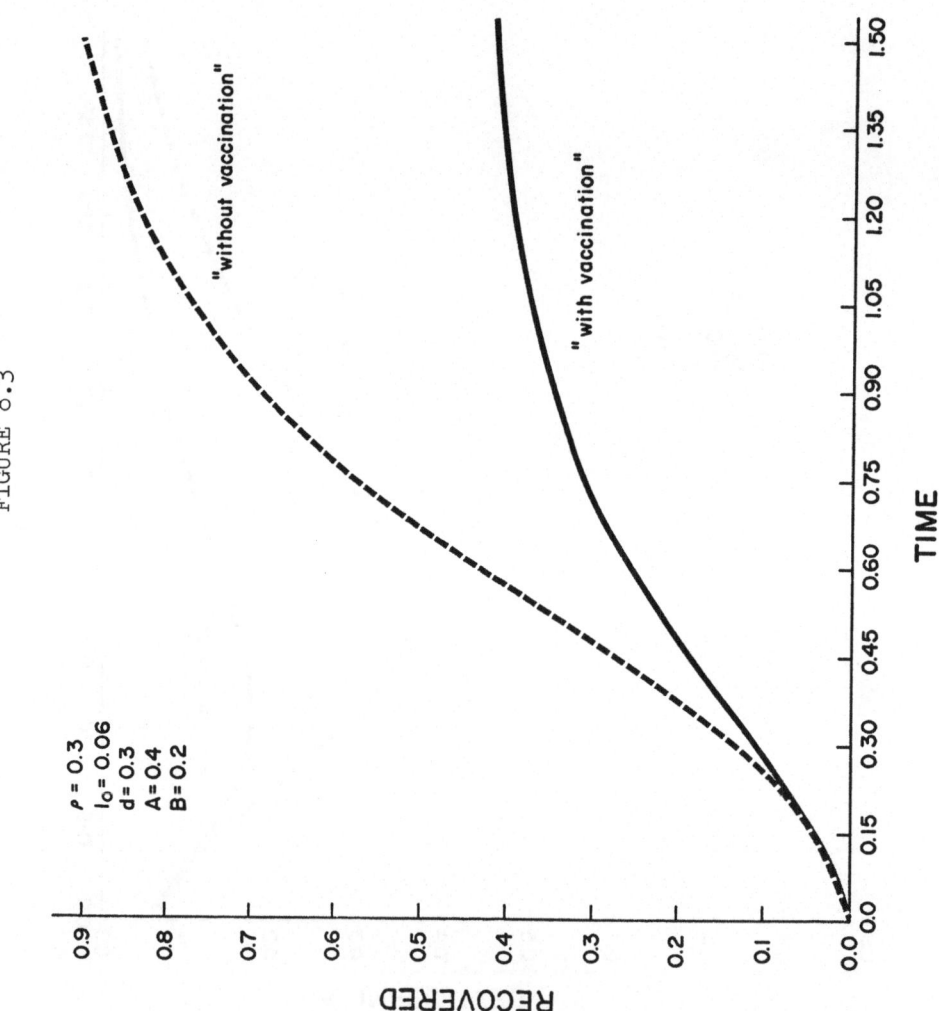

FIGURE 8.3

$\rho = 0.3$
$I_0 = 0.06$
$d = 0.3$
$A = 0.4$
$B = 0.2$

"without vaccination"

"with vaccination"

RECOVERED

TIME

FIGURE 8.4

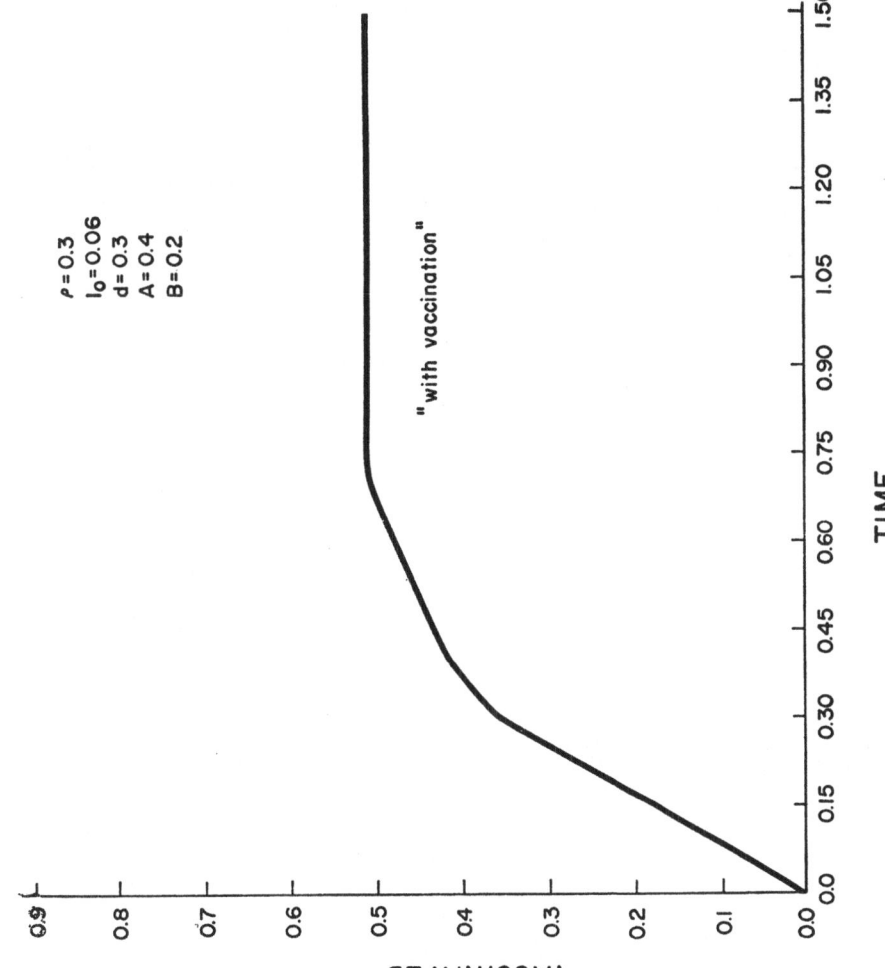

$\rho = 0.3$
$I_0 = 0.06$
$d = 0.3$
$A = 0.4$
$B = 0.2$

"with vaccination"

TIME

VACCINATED

FIGURE 8.5

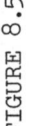

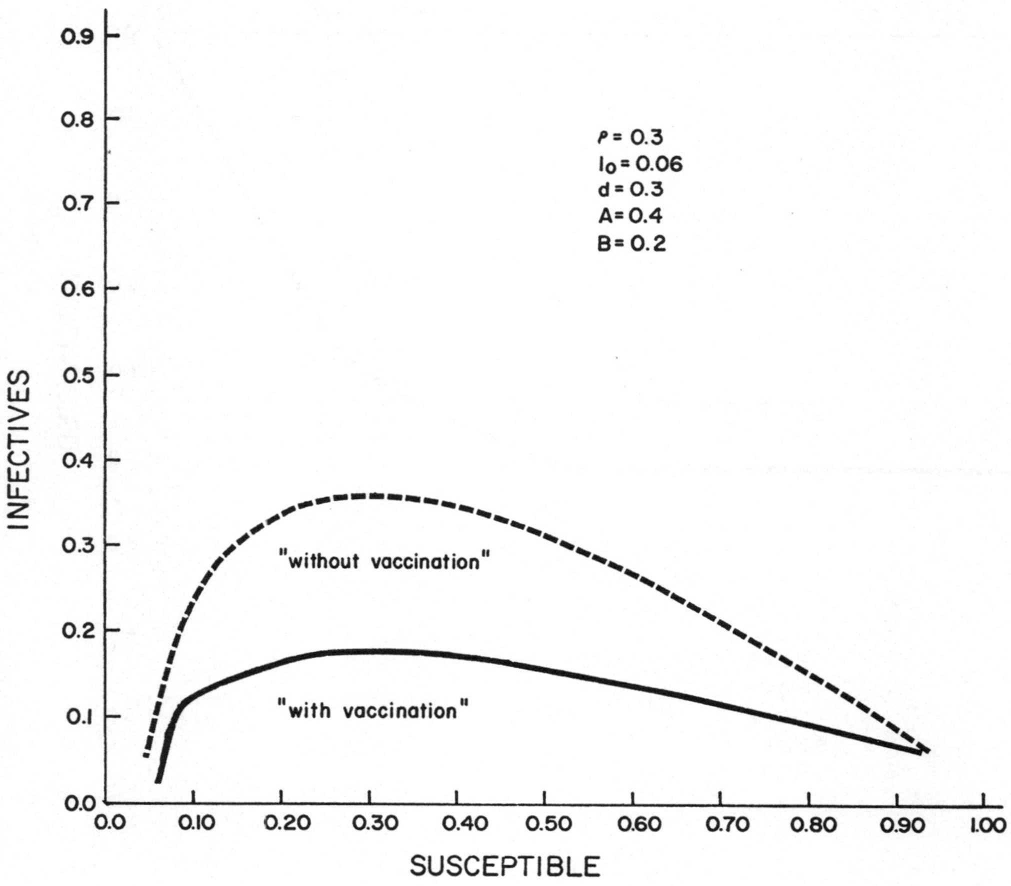
ρ = 0.3
I₀ = 0.06
d = 0.3
A = 0.4
B = 0.2
"without vaccination"
"with vaccination"
INFECTIVES
0.0
0.1
0.2
0.3
0.4
0.5
0.6
0.7
0.8
0.9
SUSCEPTIBLE
0.0
0.10
0.20
0.30
0.40
0.50
0.60
0.70
0.80
0.90
1.00

M. S. Tierney, P. Waltman and G. M. Wing, Some
Problems in the Optimal Design of Shields and
Reflectors in Wave and Particle Physics, SIAM
Rev. 7 (1965), 110-113.

Subsequently they have been used in

H. Hethcote and P. Waltman, Optimal Fractionation
Schedules for Radiation Therapy, Radiation Research, 56(1973), 150-161.

Other problems in the control of epidemics may be found in

J. L. Sanders, Qualitative Guidelines for Communicable Disease Control Programs, Biometrics 27
(1971), 883-893.

H. M. Taylor, Some Models in Epidemic Control, Math.
Biosciences 3(1968), 383-398.

D. L. Jaquette, A Stochastic Model for the Optimum
Control of Epidemics and Pest Populations, Math.
Biosciences 8(1970), 343-354.

W. K. Gupta and R. E. Rink, A Model for Communicable
Disease Control, Proc. 24th Annual Conf. on Engrg.
in Biology, 13(1971), 296.

Simulation techniques were used in

L. R. Elveback, E. Ackerman, G. Young, et.al., A
Stochastic Model for Competition between Viral
Agents in the Presence of Interference, Amer. J.
Epidem. 87(1968), 373-384,

L. R. Elveback, E. Ackerman, L. Gatewood, et.al.,
Stochastic Two-Agent Epidemic Simulation Models
for a Community of Families, Amer. J. Epidem.
93(1971), 267-280.

to determine vaccination schedules.

A new journal starting in 1974

Journal of Mathematical Biology

Editors: H.J. Bremermann; F.A. Dodge; K.P. Hadeler

After a period of spectacular progress in pure mathematics, many mathematicians are now eager to apply their tools and skills to biological questions. Neurobiology, morphogenesis, chemical biodynamics and ecology present profound challenges. The **Journal of Mathematical Biology** is designed to initiate and promote the cooperation between mathematicians and biologists. Complex coupled systems at all levels of quantitative biology, from the interaction of molecules in biochemistry to the interaction of species in ecology, have certain structural similarities. Therefore theoretical advances in one field may be transferable to another and an interdisciplinary journal is justified.

Subscription information upon request

Co-publication Springer-Verlag Wien · New York —
Springer-Verlag Berlin · Heidelberg · New York.
Distributed for FRG, West-Berlin and GDR by Springer-Verlag Berlin · Heidelberg.
Other markets Springer-Verlag Wien.

Springer-Verlag
Berlin Heidelberg New York
München Johannesburg London Madrid New Delhi
Paris Rio de Janeiro Sydney Tokyo Utrecht Wien